# BUY OR BUILD?

**The
Best
House
for
You**

*Other Books by Laurence E. Reiner*
METHODS AND MATERIALS OF CONSTRUCTION
HANDBOOK OF CONSTRUCTION MANAGEMENT

# BUY OR BUILD?

*THE
BEST
HOUSE
FOR
YOU*

Laurence E. Reiner

Prentice-Hall, Inc., Englewood Cliffs, N.J.

Buy or Build? The Best House for You
by Laurence E. Reiner
Copyright © 1973 by Prentice-Hall, Inc.
All rights reserved. No part of this book may be
reproduced in any form or by any means, except
for the inclusion of brief quotations in a review,
without permission in writing from the publisher.
Printed in the United States of America
Prentice-Hall International, Inc., London
Prentice-Hall of Australia, Pty. Ltd., North Sydney
Prentice-Hall of Canada, Ltd., Toronto
Prentice-Hall of India Private Ltd., New Delhi
Prentice-Hall of Japan, Inc., Tokyo

10 9 8 7 6 5 4 3 2

Library of Congress Cataloging in Publication Data
Reiner, Laurence E
Buy or build?
  1. House buying.    2. House construction.
I. Title.
TH4817.5.R45      643      72-11697
ISBN-0-13-109314-2

## FOREWORD

This book has been written from a background of over forty years of experience in planning, building, and managing all kinds of housing.

It points out the advantages of owning your own home, whether you plan and build it yourself or purchase one already built; whether it be a modest FHA-financed dwelling or the latest in suburban splendor. It stresses the importance of a sympathetic relationship between owner, architect, and builder, and points out the things you should know if your house is to be convenient, attractive, and suitable to your life-style. Emphasis is placed on the importance of quality, workmanship, and material. A slightly higher initial cost can reduce maintenance expense and eliminate future worries.

The book contains many cautions against the mistakes that can be made by the inexperienced and the unwary. Your home is one of the larger investments you will make in your lifetime, and you must at all times know exactly what you are doing. Used as a guide, this book can help you buy, build, maintain, and enjoy the best house for you.

Laurence E. Reiner

ACKNOWLEDGMENTS

I am grateful for the help I have received from many sources, in particular Mrs. Elisha Keeler, Reference Librarian of the Darien Library. I would also like to thank my proofreader and editorial assistant for refining and clarifying the text, and my wife, Karin, who typed the manuscript and to whom this book is dedicated.

# CONTENTS

|     |                                                        |     |
| --- | ------------------------------------------------------ | --- |
|     | List of Illustrations                                  | xi  |
| 1.  | Before You Start—Your Budget and Your Life Style       | 1   |
| 2.  | Choosing Your General Locality                         | 9   |
| 3.  | The Neighborhood and Your Site                         | 21  |
| 4.  | Choosing the House That Is Best for You                | 35  |
| 5.  | The Development House                                  | 49  |
| 6.  | The Speculative House Built by a Local Builder         | 67  |
| 7.  | Buying a House That Has Been Lived In                  | 77  |
| 8.  | Rehabilitating an Old House                            | 85  |
| 9.  | Developments on Artificial Lakes—and Vacation Homes    | 93  |
| 10. | Architecture and the Livable House                     | 101 |
| 11. | The Heart of Your House: Work and Utility Areas        | 123 |
| 12. | Alteration and Remodeling                              | 137 |
| 13. | Building Your Own House                                | 147 |
| 14. | Insurance                                              | 171 |
| 15. | Mortgages                                              | 177 |
| 16. | Basic Real Estate Law                                  | 187 |
| 17. | Landscaping                                            | 195 |
| 18. | Room for Recreation, Hobbies, and Sports               | 205 |
| 19. | Maintaining Your House                                 | 215 |
| 20. | Fire Alarms and Burglar Proofing                       | 229 |
|     | Index                                                  | 235 |

# List of Illustrations

Figure 1: An Example of Cluster Housing     43
Figure 2: Section of Foundation for Flat Concrete Slab     56
Figure 3: Section of Foundation Walls and Footings     57
Figure 4: A Structural Frame for a Typical Wood House     58
Figure 5: Types of Gutters     60
Figure 6: Flashing     71
Figure 7: Suggested Electrical Outlets for a Living Room     74
Figure 7A: Suggested Electrical Outlets for a Bedroom     75
Figure 8: The Split-Level House     105
Figure 9: The Mansard Roof     108
Figure 10: Ways of Entering a House from Garage to Work Area     116
Figure 11: Ways of Entering a House from Front Entrance without Going Through Length of Living Room (in Absence of Center Hall)     117
Figure 12: An 8 by 12 Foot Kitchen     126
Figure 13: A 7 by 9 Foot Bathroom     130
Figure 14: A 2 Foot 6 Inch by 5 Foot Powder Room     132
Figure 15: A Combination Laundry, Mud Room, and Pantry     133
Figure 16: A Cavity Wall     163
Figure 17: Land Plan for a Small Plot     199
Figure 18: An Example of an 8 by 10 Foot Workshop     210
Figure 19: A Shop for Automobile Repair and Restoration     211
Figure 20: Firestopping a Wood-Framed House     223

# CHAPTER 1

## BEFORE YOU START — YOUR BUDGET AND YOUR LIFE STYLE

1. Before You Start to Look
2. Examine Your Mode of Living
3. How Large a House Do You Want?
   A. Your Cash Position
   B. Your Budget
4. Your Future Cash Position and Prospects

## 1. BEFORE YOU START TO LOOK

If you are a typical American family, at some point in your life you will go house-hunting. Perhaps you have outgrown your house, or your house has outgrown you. Perhaps your job has called you to another area. Or perhaps you finally have attained that financial plateau which enables you to buy and live in a house of your own.

The instinct for a home is a natural one. Nothing makes a man feel more secure than his own roof over his head, no matter how mortgaged. Do give way to the irresistible urge to dig in your own dirt, decorate your own rooms, look out your own windows. But please be careful! Don't be stampeded by pretty pictures or glib real estate agents or irresponsible developers. Above all, don't expect to find the Perfect Home.

If you are *buying*, you must be prepared to compromise—between a magnificent old oak in the yard and ancient plumbing; a modern kitchen and no basement; charming low ceilings and a 6'4" husband. Neither discount nor be seduced by some feature that may delight you at first glance; it can prove to be a source of exasperation as well as delight. By weighing practical considerations against appearances, it may be found that the virtues will outweigh the disadvantages.

If you are *building* your first house, you also must be prepared to compromise—between a view of a brook and a western exposure; proximity to schools and a noisy thoroughfare. You will need to make many choices as you prepare to buy or build a house. You will adjust and learn to accept, and although you may not have the Perfect House, in time you will be happy with the Perfect Home. This book is meant to help you make these choices as painlessly and with as few errors as possible.

Before you start to look, be sure you know what style of house you want, where you want to live, and what you can afford. These three items are so important that I have approached them from several different angles to be sure every facet is covered, every pitfall explored. I beg the reader to be patient with what may sometimes seem like unnecessary repetition. The following pages will provide the all-important overall look that can prevent longstanding disappointment or the financial difficulties that come with unforeseen extra costs. In extreme cases, your whole investment may be lost because of inferior construction or a bad location.

## 2. EXAMINE YOUR MODE OF LIVING

This is a country of upward mobility. The butcher, the baker, the candlestick maker can now hope to send their children to college and to have a house of their own. Their children, many of whom have followed in their father's footsteps and many of whom have entered professions, also wish to have a house of their own. The question is: what kind and where? To answer intelligently, you must first examine your present mode of living and your life style.

It is important to think of where you will be comfortable and happy. Would you rather live in a development where houses are priced alike and where you will live with people of your general income and station in life? Or would you prefer an older established neighborhood where getting acquainted is not so easy and where there may be wide variations in income and status? If you now live in a large city and your work is located in the city center which you can reach by public transportation, you must decide whether you wish to continue using such transportation or are willing to drive to work every day. If you attend school after work or engage in other after-hour activities, it may be important to live in a location which can be conveniently reached from these activities. If you have school-age children, you must think of where the good schools are. You should consider your general interests and hobbies, such as gardening, golf, swimming, or just plain loafing in your own back yard with a glass of beer. Are you more of an indoors-lurker or an outdoors-lover? Are you partial to the casual and convenient—or to the stylish and imposing?

You also must consider your future prospects. If you are planning to look for a house that will be only temporary until you can afford a better one, this must be taken into account. If you enjoy working with your hands you may want to look for a house that is reasonably priced because it requires a great deal of repair and modernizing. Many young couples do this with varying degrees of success. It is not easy to rebuild a house or make major repairs after finishing a day's work. Do you have the energy and determination to carry this through?

If you have special ideas about the kind of house you want, you might prefer to build one. If you have the means, do it! In this case you will need to look for vacant land and for an architect and contractor. All of these subjects will be dealt with in detail in later chapters. First we must come to terms with the basic mundane factors from which all other house-buying and house-building decisions proceed.

## 3. HOW LARGE A HOUSE DO YOU WANT?

Your mode of living and the size of your family are factors in determining the size of the house in which you can be comfortable, but remember that unless you are extremely wealthy you cannot afford everything you would like to have and you must be willing to make some compromises. Consider the number of rooms you require as well as their size and layout. If you have several children you may have to put at least two of the same sex in a single bedroom until you can afford more room. Perhaps you can buy or build a house with an attic and dormers for future expansion, or one laid out in such a way that extra rooms may be added later. Every house, of course, has a kitchen and at least one bathroom but there is a wide variation in their sizes and equipment. Perhaps at the start you can plan to eat in the kitchen with your family and use the living room for party buffets. A single bathroom can be extremely inconvenient if parents and the children all have to leave for work and school at the same time. Adding extra bathrooms to a finished house is expensive, however, and a compromise could be a "powder room" containing a basin and toilet bowl. If you have a hobby requiring work space, look for a house with a usable basement or attic. You may want more land than is generally available with a house, or you may be content with a 50 × 100 foot plot on which you can, if you are ingenious, do extensive gardening. Add up these requirements before you start looking. To summarize: you need a living room; a kitchen; a master bedroom; $X$ number of bedrooms for the children; a guest bedroom, if you plan to use it often (otherwise it is an expensive luxury); at least 1½ bathrooms; space for hobbies or family activities; and enough land for your needs or taste. (A sewing room, library, or study may be used temporarily as an extra bedroom by purchasing a convertible sofa. Do you have many guests—often?)

A.  YOUR CASH POSITION

In addition to examining your mode of living and the size of the house you need, you must carefully consider your cash position and your budget. Purchasing a house takes cash. No matter what kind of mortgage you are able to obtain and how much of the cost of the house the mortgage covers, you must still pay for such things as title searches, attorney's fees, mortgage closing costs, and, of course, the difference between the amount of the mortgage and the price of the house. In addition to the actual cost of the house, you certainly will want to spend some money for planting, rugs, draperies, and other amenities to change your house to a home. You must, therefore, check your present cash position to make sure that you will have enough money on hand to meet all first costs and still have a little left over for the "cosmetics."

### B. YOUR BUDGET

In addition to your present cash position, consider the monthly carrying costs of the house you have in mind. In order to arrive at the amount you can afford for such "rent" you will have to prepare a careful budget of your entire income and outgo. You must consider the fact that your new rent will contain items which you probably did not have to pay as a tenant, such as heating fuel, gas, electricity, insurance, and general maintenance of the premises. In examining your budget, first enter all your essential expenses except rent—these include such items as food, clothing, utilities which you now pay for, medical expenses, education, union or other important dues, transportation, recreation, vacations, and so forth. When you subtract these costs from your income, you will obtain a figure which should determine how much money you will have available for "carrying" your house. The checklist at the end of this chapter will guide you in the preparation of your budget as well as in spelling out the expenses you are likely to face when you own your own house.

## 4. YOUR FUTURE CASH POSITION AND PROSPECTS

Everyone hopes to better himself and you must think of this when you are considering your new house. The entire idea of a house for yourself is colored by your hopes for the future, and by what you want for yourself and your children. This does not mean shelter only. You should think about your children's schooling, the people you want them to meet, and the children you want them to play with. Think of the people you want to live near and the people with whom you will be comfortable. You should consider your future job prospects, your future family size, and the amount of room you will require now and later. If you are optimistic regarding your prospects for

# BEFORE YOU START

advancement, it may do no harm for you to reach a little further than you can now comfortably afford, provided, however, that you always have something to fall back on in the event of an emergency. (This can mean one or more of many things—money in the bank, an understanding employer, or a rich relative!) A house in a good neighborhood with extra room for future members of the family is a very nice thing to own and can be a source of present and future pleasure. But if you are one of the many families who have to strain every resource in order to buy a house, do *not* buy a house that is larger than you can afford. You can always buy a larger one when your financial position improves.

Remember that your house probably will be one of the largest investments you will ever make and you must insure its security. Consider your life insurance, health insurance, and any investments you may have that can be used to keep your house safe for you and your family. No one can be certain of the future but the prudent man can prepare for it wisely.

### CHECK LIST

*Your Monthly Budget*

1) Food
2) Allowance for lunches, etc.
3) Fares (commutation, buses, etc.)
4) Electricity and Gas
5) Telephone
6) Insurance
7) Medical—Dental
8) Automotive (including all automobile-related expenses—gas, oil, tires, repair, insurance, garaging, payments or rental)

9) Clothing
10) Dry cleaning, laundry, clothing repairs
11) Household furnishings and appliances
12) Education (tuition or lesson fees)
13) Entertainment
14) Contributions
15) Dues and memberships
16) Monthly installments (except for items which may be included in one of the previous categories)
17) Miscellaneous

Deduct the total of these from your monthly income, after taxes, and you will have the sum available for "rent" or monthly payments on your house. These payments will include:

1) Mortgage interest and amortization
2) Fuel for heating and cooking, water, garbage removal
3) Insurance on your house
4) Taxes (which may be included in your monthly payment to the bank or other lender)
5) Repairs and maintenance

CHAPTER *2*

# CHOOSING YOUR GENERAL LOCALITY

1. The Available Choices
   A. The City
   B. The Suburbs
   C. The Country (Exurbs)
   D. The Complete New City
2. The Elements to be Considered in Making a Choice
   A. Transportation
   B. The Economic Level of the Locality
   C. Schools
   D. Taxes
   E. The Conveniences of the Community
   F. Zoning

1. THE AVAILABLE CHOICES

As mentioned in Chapter 1, there are many reasons why people are in the market for a house. They may have been transferred from another city; they may be tired or unhappy with the present neighborhood or with apartment living; they may be newlyweds; a promotion may have inspired the move to a better house; or they may simply want a home of their own. People who are looking for a house to settle in come from widely varied social and economic backgrounds, and their choices of locality usually reflect these differences.

The choices available to you depend on the size of the city or town in which you live or work or to which you have been transferred. The smaller the locality, the narrower the choice; indeed, if the town is small and is dominated by your company there may be no choice at all. But this is rare and normally there are several areas where you can locate.

A. THE CITY

If you start with the city proper you will find that all cities have either begun or are in the late planning stage toward the revival of the central city. There are many opportunities for the house-buyer in such areas. Every large city also has semi-suburban areas within the city proper, which have all utilities in place as well as available public transportation.

Even a small city or large town has outlying areas within its corporate limits, and if you want a larger piece of land than is usually available close to the center you may have to go there

B. THE SUBURBS

We now come to the actual suburbs. These are areas outside of the corporate limits of the cities and they have governments

CHOOSING YOUR GENERAL LOCALITY 11

of their own. They have their own school systems and levy their own taxes. You may have as many as a dozen or more suburban towns to choose from. Some of these towns (or villages) will have housing available for sale in almost any reasonable price bracket.

C. THE COUNTRY (EXURBS)

If you want to live in a rural area with a good deal of land of your own you usually can find such areas within thirty to forty miles of even the largest urban centers.[1] In some of these rural areas, land speculators have purchased large tracts which they have divided into building lots (sometimes of several acres each) on a plan which also shows future roads, utilities, and so forth. These "paper subdivisions" exist only on a plan and the land itself is still in its natural condition.[2] You may purchase such land if you wish to improve it and build on it or you may purchase a house with several acres if you wish to buy a completed project. Some subdivisions have been developed by the owner with roads and some utilities. Others have been developed further with houses ready for occupancy. Some of these rural areas (it has become fashionable to call them exurbs instead of suburbs) are surprisingly close, in point of time, to central cities.

---

[1] Further out than this is not recommended if you are planning to work in the central city.

[2] A subdivision may simply consist of a plan showing a tract of land which has been divided into building lots showing the future location of road, utilities, etc. The entire tract may be covered with heavy brush or trees or it may be rocky or swampy; anyone who buys such a piece of land must clear the land, and provide access and utilities at his own expense.

### D. THE COMPLETE NEW CITY

Another choice may be a completely developed subdivision which is large enough to provide within itself a school, church, shopping facilities, recreational facilities, and even direct transportation to nearby urban centers. The entirely new planned small city as a satellite of a large urban center is gaining a foothold in England and in Scandinavia. Unfortunately, it has not yet taken hold in this country and very few examples have been planned or built. How will you choose among them or between the suburb and the city proper? The following sections may help.

## 2. THE ELEMENTS TO BE CONSIDERED IN MAKING A CHOICE

Your economic bracket and mode of living will, of course, influence your choice of the general locality in which you want to live, but even within such limits there are still many choices.

### A. TRANSPORTATION

How will you get to work? Some major cities still have a viable network of commuter railroads leading into the central city. Other cities have freeway or superhighway networks that reach into their centers; many with such highway systems have very poor public transportation so that your only choice is to drive. Driving to a central city is a daily horror unless you have steady nerves and an iron constitution. Public transportation leaves a great deal to be desired but at least one can read or take a nap. Many cities are developing high-speed bus lines which very often go directly to outlying districts with no intermediate stops. If you work or expect to work in the central city you should investigate the available public transportation, the outlying areas it serves, how reliable it is, and how long it takes to get to the city. Such information is available from the urban Chamber of Commerce or the various community chambers. In any case, you are advised that one hour on train, bus, or private car is sufficient. If you allow one-half hour more from portal to portal, you have three hours a day of travel time. If you work for a company located in a suburb you may have a choice of many localities within one hour of driving time from house to office.

If you prefer, there are available areas on the fringes or within the central city. The advantage of being home within an hour after leaving work and of not having to drive because of good public transportation is worth a great deal to many people. Perhaps the checklist at the end of this chapter will help you make up your mind.

B. THE ECONOMIC LEVEL OF THE LOCALITY

Once you have located the communities with satisfactory transportation to your job, it is time to consider—and this is a very vital factor—how much you can afford to spend for a house. There are suburbs or even semi-suburbs within the city limits where housing may not be available at prices you can afford to pay. Many suburban communities have restrictive zoning codes against houses which are under certain heights or floor areas. Some communities restrict development by establishing building lot areas which are larger than speculative builders can sustain economically. These restrictions affect you, also, if you wish to build.

You should study the real estate sections of newspapers for house prices; drive through selected communities to get an idea of how the people live and the size of the houses. It is also advisable for you to consult real estate agents in these communities to tell them your requirements and obtain their reaction. You should also speak to your fellow workers and friends.

## C. SCHOOLS

The level of schooling varies among communities and the wise parent will look into this very carefully. There are plenty of schools that will give the student an excellent basic education in reading, writing, and arithmetic and which will prepare him for well-paying and thoroughly respectable occupations, which do not require a college degree. Such schools exist in localities where the economic and social status of the population requires them. There are other communities which have a reputation for the number of their high-school graduates who go on to college; schools in these communities have extensive courses in art, music, languages, higher mathematics, and the sciences. The taxes to support these schools make up a large proportion of the localities' tax roll. If your children are grown, or you have none, you are advised to avoid communities with a high reputation for the excellence of their schools unless you are willing to pay the taxes that go with such a reputation.

In referring to schools and the level of schooling in the various communities, it is only fair to call to the reader's attention the attempts that are being made in many sections of the country to change the method of taxation for schools, and indeed, to do away with corporate school boundaries. The reader must bear this in mind, and based on the best available local information, judge for himself what effect such laws may have on the local schools.

## D. TAXES

Investigate the tax rate of your chosen localities. Don't go only by the mill rate, which is the number of dollars you will pay per $1,000 of assessed valuation.[3] (Using the mill rate, for instance, if the tax rate is 40 mills, then the tax on a house assessed at $10,000 will be ten times $40 or $400.) *You must*

---

[3] A mill is 1/10 of a cent.

*look at the rate of assessment in order to obtain the true tax rate.* Thus, if the locality assesses at 50 percent of value, then the 40-mill rate of the $10,000 assessment on a $20,000 house is actually only 20 mills on the true value. The 20 mills is the true rate (50 percent of 40 mills). If the assessment rate is 80 percent of value, then the true rate is 80 percent of 40 mills, or 32 mills, which means that on a $20,000 house you would pay $20,000 times 80 percent times $40 per thousand, or $640. Some states assess at 100 percent of the market value of the house, in which case your tax would be $800. Sometimes tax rates are quoted in dollars, which means so many dollars per $100 or per $1,000 of assessed valuation. Again, you must know the rate of assessment to find out what your taxes really will be.

When you are investigating the current tax you should prepare yourself against unpleasant tax surprises. Check the history of the assessment rate and the mill rate. The local tax assessor at the Town Hall will be glad to help you. The best you can expect is that the mill rate has kept pace with the cost of living. Some towns, however, may have undertaken a heavy load of capital expenditures, such as new schools, sewers, public buildings, or new equipment, and in these places the tax rate will show a rapid rise. Also check the history of the assessment rate. Find out how often the town reassesses. Find out whether they immediately reassess a newly purchased house or whether they will leave it until the next assessment time. (They may assess it still higher because of its sales price.)

Spend some time on the tax situation. Once you buy or build a house, you are there and your mobility is limited. Taxes can be very burdensome.

## E. THE CONVENIENCES OF THE COMMUNITY

The conveniences or amenities of the locality—an attractive, easily reached shopping center, local specialty shops that sell everything from baby clothes to antiques—are important to your well-being and pleasure. Enterprises which cater to your hobbies may be a determining factor. A nearby golf club, public or private, that you can afford, a bowling alley or tennis courts, are important to many people. A town situated near the water is extremely provocative for the determined sailor or fisherman.

Many people look for cultural pursuits. A good local library, proximity to museums, local orchestra and theater groups, can be persuasive for many people. Other things being equal, it is very possible that you may choose the community which caters to your interests, whether physical, intellectual, or both.

## F. ZONING

The zoning laws and the enforcement of the zoning code in a community are crucial factors in your choice of the area in which you wish to locate your home. Zoning is as important as taxes, schools, or any other subject mentioned in the preceding sections. If the zoning code is badly drawn and loosely administered, it can reduce the real estate values and undermine the way of life of the community. The purpose of a zoning code can be best expressed by the following excerpts from the preamble to a typical zoning ordinance.

1) To promote and to protect the public health, safety, morals, comfort, convenience, and the general welfare of the people.

2) To divide the [city, town] into zones and districts restricting and regulating therein the location, construction, reconstruction, alteration, and use of building structures, and land for residence, business, commercial, manufacturing, and other specified uses.

3) To protect the character and maintain the stability of residential areas ... and to promote the orderly and beneficial development of such areas.

4) To regulate the intensity of use of zoning lots ... necessary to provide adequate light and air. ...

## CHOOSING YOUR GENERAL LOCALITY

5) To establish building lines and the location of buildings designed for residential use.

6) To prohibit uses... which are incompatible with the character of the area.

7) To conserve the taxable values of land and buildings throughout the [city, town].

There are many other reasons for zoning ordinances, but the above refer specifically to residential areas. The reader will note that, among other things, a municipality is interested in conserving the taxable value of property. When it protects the residential taxable values the locality is also protecting your values.

There is no mystery about zoning, and every municipality will give or sell you a zoning map and code so that you can see exactly where the various residential zones are located; what restrictions there are, such as minimum and maximum lot sizes; how the residential zones are located with relation to commercial and industrial uses; and whether there has been an influx of multi-residential building into zones where most of the houses are still privately owned.

You can also see where development housing is permitted with relation to commercial and industrial use.

One other important matter with relation to zoning: Many towns are under pressure from commercial enterprises, real estate developers, apartment house builders, and other land speculators to break down the zoning ordinances in order to allow nonconforming uses in residential areas and to lower the requirements for the sizes of building lots allowed in the various zones. You can obtain an idea of the extent to which the town has given way to such pressure and of how strictly it enforces the present code by speaking to the local real estate brokers, or by asking the enforcing official what zone changes have been made recently, have been requested, or are about to be made.[4]

---

[4] The enforcing official may also be the local building inspector or he may report to the Planning and Zoning Board. You can find him at the local town hall.

Don't forget how important this zoning matter is! It can drastically affect both your investment and your satisfaction with your new home.

## CHECK LISTS

*Transportation*

1) Commuter Railroad

   a) Travel time to your station—how will you get there?
   b) The distance from terminal to your place of business.
   c) Cost, including local fares.

2) Bus Line (express or local) or subway

   Same points as 1a, 1b, and 1c.

3) Expressways

   a) Time to place of business.
   b) The available parking and how much it costs.
   c) The desirability of owning two automobiles.
   d) Total cost of driving—insurance, depreciation, tolls, parking, gas and oil, car-pool fees.

4) Other ways

*The Economic and Social Level*

1) Does the locality have sharply defined neighborhoods?
2) Look at the people and their homes. Is this

where you want to live? Will you be happy here?

3) Get out of your car and walk around the shopping center or the local main street.

4) The houses available in your economic bracket. Don't look yet—obtain a general listing from a real estate broker first.

*Schools and Taxes (These are combined because the schools get the bulk of your tax dollar.)*

1) The general reputation of the educational system.

2) The percentage of high-school graduates who go to college. The local Board of Education office will know.

3) Take time during school hours to get a look at the student body, especially at the high school.

4) The present true tax rate.

5) Check the history of this tax rate.

6) The frequency of property reassessment.

7) The locality's plans for new public buildings and other public utilities. The local newspaper or a real estate broker will know.

*The Amenities and Your Hobbies*

Shopping centers

Specialty stores

Churches

Libraries

Cultural pursuits

Golf courses

Bowling alleys

Tennis courts

Salt or or fresh water for sailing, swimming, or fishing

After you have checked the items on these lists you must weigh the facts against your own preferences. Perhaps you are willing to suffer bad transportation to your job if your children can walk to a good school. If you love golf and have grown children, or no children at all, perhaps you want to live in an older neighborhood near a good public or private golf course and are willing to suffer some inconvenience to do this. You will rarely find an ideal situation, but you can settle for less and still be happy.

# CHAPTER 3

# THE NEIGHBORHOOD AND YOUR SITE

1. The General Impression
   A. The City Renewal Area
   B. Suburban Areas
   C. Condominiums and Cooperatives
   D. The Country House
2. The Quality of the Particular Site
   A. Transportation
      1) Getting to Work
      2) Getting to Shopping and Schools
   B. Zoning
      1) The Zoning of Your Location
      2) Neighborhoods with Older Houses
      3) Condominiums and Cooperatives
      4) Mobile and Modular Houses
   C. Utilities
      1) Water and Sewer Lines
      2) The Town Roads
   D. Topography
      1) Sun Direction
      2) The Lie of the Land
      3) Danger of Mudslides or Landslides
      4) Trees
      5) The Elevation
   E. Is Your Land on an Island?
   F. Prevailing Winds
   G. Riparian Rights
   H. A Final Thought on the Location

## 1. THE GENERAL IMPRESSION

When you have chosen the locality where you think you will be happy and comfortable, it is time to take the next step. You now will have to start looking at particular houses or building lots in specific neighborhoods. You may be looking for a city house or an old farmhouse, but the basic requirements for pleasant living are the same.

When you go to look at a particular house or building lot, you must take enough time to look around carefully and to gather impressions of what it would be like to live there. This is the *ambiance* of the area.

### A. THE CITY RENEWAL AREA

If you are looking for a city house in a neighborhood that is being rehabilitated, don't expect too much. You certainly will find a mixture of old dilapidated houses, houses in the process of being done over, and completely renovated houses. You may find a number of the former inhabitants still about and you must not forget that you are in a formerly depressed area. Drive slowly, or better still, walk around the neighborhood. Examine the renovated houses to gain an impression of how well they have been done. Try to get an idea of what the new people are like and how long it will be before the area will become stabilized.

### B. SUBURBAN AREAS

If you are looking for a house in an established neighborhood in the suburbs, you should try to see it on a weekday as well as on a weekend. On weekends (in the spring, summer, or fall) the husbands usually are doing yard work and the children probably are playing. (What kinds of games are they playing, and how do

# THE NEIGHBORHOOD AND YOUR SITE 23

they compare in age to your children?) Or couples may be entertaining. During the week, housewives may be hanging out the wash, going shopping, or driving the children to school. As you will be living there seven days a week, you should get a complete impression. This is also true if you are interested in a house in a new development.

Many developments or housing estates recently have been hacked out of farmland or wooded areas and give a feeling of rawness. You should not be discouraged by this if you like the feel of the area. Trees and lawns and flowers will grow and the area can get to look established in a very few years.

## C. CONDOMINIUMS AND COOPERATIVES

The same general rules are valid if you are looking for a condominium, a cooperative, or a townhouse. In these instances you will be living very close to your neighbors and it is even more important that you know what kind of people they will be. In a townhouse you do have your own small backyard, but in a condominium or cooperative you have only your own apartment and must share all common facilities. Walk through the entire area. Is it a pleasant neighborhood and do the people seem friendly?

## D. THE COUNTRY HOUSE

The old farmhouse or the small exurban estate should be seen at night as well as during the daylight hours. Look for the nearest lights and listen for the sounds of traffic. Perhaps you want to be completely away from everything, but most people like to see a friendly light at night and to get at least a glimpse of a nearby house in the daylight. If you are locating somewhere in the North, think how it will look after a heavy snowfall.

To repeat: before you make any commitments, spend some

time near the house you would like to buy to see if you will feel comfortable living in it.

## 2. THE QUALITY OF THE PARTICULAR SITE

The quality or character of the particular building lot or house in which you are interested has many variables. You must choose among them to decide which you can live with and which could annoy you to the point of discomfort in your daily living. By all means *try* to find the perfect site, but be willing to compromise.

### A. TRANSPORTATION

1. Getting to Work

Transportation has been mentioned in general several times before, but now we must consider it in terms of a particular site. If commutation to the husband's place of business is by railroad or intercity bus, his means of transportation to the bus line or the railroad station must be considered. A half-mile walk can be fine in good weather, but on rainy, snowy, very cold, or very hot days he may have to be driven. If there are small children and only one car this can become difficult; if a second car is necessary the cost must be considered. If the husband drives to work and you are a one-car family, you must, of course, find a house or lot location from which all the facilities required for daily living can be reached by walking or by public transportation. (See the checklist at the end of this chapter.)

2. Getting to Shopping and Schools

The wife also must get to shopping areas and get the children to school. The local school bus comes under the heading of transportation; you would be wise to find out if your children

# THE NEIGHBORHOOD AND YOUR SITE

will be eligible to ride on it. (Eligibility is determined by distance and school districts.) It is often more convenient to go into town by bus than to go through the trouble of driving and parking. A nearby local bus stop can be very handy.

## B. ZONING

Again we come from the general to the specific. Zoning is a very important item which you must investigate when you are choosing the locality where you want to live.

### 1. The Zoning of Your Location

After you have tentatively selected a house location, you must examine the town zoning map very carefully. Find your location on the map and find what zone you are in. Then see what the zone allows for sizes of building lots; front, side, and rear yard distances from the lot line; and possibly house sizes. Check then to see how your zone is located with relationship to higher or lower zoned areas. *Be careful that you are not on the edge of a commercial or industrial zone. If you are surrounded by an all-residential zone it is better to be closer to a higher use than to a lower one, particularly if the lower use allows apartment houses or multi-family dwellings.*

If you are interested in a house in a new subdivision or development, check carefully to make sure that the development has not been pushed out to the poorer fringes of the town where industry or heavy commercial use is permitted. Check especially carefully to see if the subdivision is on the edge of a large undeveloped area. Be sure such areas are zoned for residential use.

### 2. Neighborhoods with Older Houses

Older houses in small towns naturally tend to be near the town center. Make sure that retail commercial business is not beginning to encroach and be positive that there is no prospect of down-zoning the area to allow multi-family dwellings. Ask the local zoning official and the broker if there is any prospect of down-zoning. Also look for yourself. Drive and walk around. If all the houses are well-kept, with fresh paint and tended lawns and flower beds, the chances are that the area is secure—at least for the immediate future. If there is a town planner, ask to see the town plan map. He will be pleased to show it to you, and it will tell you what possible zoning changes there may be in the future.

3. Condominiums and Cooperatives

It is well to investigate the zoning even if you are interested in a multi-family project such as a condominium, a cooperative, or a townhouse. If these are of the low garden apartment type, you don't want to be near high-rise apartment buildings or too near retail shopping or other commercial or industrial use.

4. Mobile and Modular Houses

Mobile houses and certain types of modular houses are not allowed everywhere in most communities. In some towns they are not allowed anywhere in any residential zone. If you are interested in such a house, it is best to make sure that you can live in it in a reasonably good area. Inquire about mobile home parks and go to see where they are. Ask mobile home dealers about locations and speak to local zoning officials. Please don't purchase a mobile home unless you are sure of locating it in a residential area.

## C. UTILITIES

1. Water and Sewer Lines

The information which you must have about utilities is especially important if you are purchasing land for a building site. If there are no public water mains you must dig a well, which can be a chancy and very expensive proposition. If there is no public sewer system you will have to install a septic tank and a drainfield. This can mean separate dry wells for washing

# THE NEIGHBORHOOD AND YOUR SITE

machines; no garbage disposal unit; no strong caustic or drain solutions (which are bad for septic tanks); and regular pumping of your septic tank. It is not a big nuisance and many people live with it comfortably. (The author has had a satisfactory septic tank system for over thirty years.) Electricity and telephone services are available everywhere; yet if you want to be far out in the country you may have to pay for extending pole lines, or you may have to pay extra if you are set far back from an existing pole line.

Most smaller towns do not have gas mains. If you are used to cooking with gas, you will have to use either bottled gas (which is expensive) or electricity for cooking.

2. The Town Roads

Although roads are not strictly a utility, it is well to look at your roadway and the surrounding roads. Are they reasonably maintained or are they pot-holed and crumbling? In the northern part of the country you probably should inquire how good the town is about snow plowing. If you live in the part of the town that is plowed out last, it can cause a hardship when the husband can't get to work or a child can't get to school or to a doctor. Many towns have excellent road maintenance facilities. This is not a must, but it is a good plus.

## D. TOPOGRAPHY

1. Sun Direction

Topography is used here as an all-embracing term meaning the physical characteristics of the site itself. For instance, if you have your choice of several houses in a development, do you want a house with morning sunlight in your bedroom or do you want it in your living room, which will therefore be shaded in the afternoon? If you have a choice of porch or terrace locations, it would seem best to have them shaded in the afternoon inasmuch as the western sun can be quite hot in the summer.

Sun direction is important in still another context. In the

South and Southwest, people who drive to work will try to locate their homes east of their place of work so that they are not facing the rising sun in the morning or the setting sun in the evening.

2. The Lie of the Land

The slope of the land can be important. If the house is already built it should be higher than—or at least as high as—the surrounding land; if not, you may have a wet basement, especially if there is considerable slope of land toward the house. If you are going to build be especially wary of marshy or wet spots on the land and rocky outcrops. The slope of the land also may be important in sheltering your house from strong winds, either hot or cold. Some parts of the country have seasonal winds which can be quite unpleasant.

Be suspicious of wide flat areas of land near large bodies of water. They may be former marshlands which have been filled in, and such fill may be a thin cover over a deposit of river or tidal silt and marsh grass. There are many cases where houses built on such fill have sunk and finally collapsed. If you suspect that such a condition exists, please inquire at the town planner's or zoning official's office or at the building inspector's office. It is worth the trouble to find out, since many people have seen their entire investment wiped out by building a house on such land. It is safe to buy *only* if the proper town officials are aware of the condition and have had the builder take proper precautions to insure a stable foundation. This can be done by piling or spread footings (see Chapter 5).

3. Danger of Mudslides or Landslides

The subject of mudslides or landslides is of such special interest to residents of certain areas, notably the Pacific coast, that the author feels it should be discussed in detail.

In the past few years the State of California and other states have become aware of the increasing danger of such slides. Because of the hilly or mountainous terrain and the unavailability of flat land, more and more people must build on hillsides. It is

# THE NEIGHBORHOOD AND YOUR SITE

to protect such people that the authorities have promulgated and are stringently enforcing rules, which the land developer must follow in preparing his land. The regulations go further in that they prevent the developer from selling his land unless he complies with them in every detail and they make him financially responsible for any future mishaps.

Coming under the heading of excavation and grading, these rules refer to any land which has been made or terraced out of a slope; and the pitch of the slope is defined for various areas. The land developer not only must file for such work and obtain permits to sell the land, but must also tell the purchaser the cost of his house foundations; these cost figures must be supported by an engineer's certificate. The regulations also define the slopes permitted for driveways and for streets. They refer to retaining walls, compaction of soil, and proper drainage.

Many lenders now require certification by engineers and governmental bodies as to the stability of the land and the adequacy of the foundation before they will consider mortgage money. If you are interested in a hillside site, you must be sure to consult with the town or county engineer or building authority, as well as with prospective lenders, in order to receive their assurance that the site is safe and that you can obtain mortgage money.

4. Trees

If there are trees on the property, look at them carefully. Old trees with large branches overhanging a roof can be dangerous during wind or ice storms. In addition, leaves in house gutters can be a nuisance. It is nice to have trees on your land; but the house should be set away from them so that you may enjoy them but not be bothered by masses of fallen buds in the spring and masses of fallen leaves in the autumn. Except in really hot climates, the shade of trees located directly over a house is not necessary. A bright sunny house is much preferred by most people. If you are buying a building lot, a later section of this book will deal with site planning with reference to trees, slopes, and so forth.

In mentioning trees it should also be noted that there are many regions that are exposed to brush fires or forest fires. If you are proposing to build or buy in one of these areas you must have a wide clearing for your house *and* a plentiful source of water.

5.  The Elevation

The elevation of the land is important. If you are on a hill in a northern part of the country you may be swept by cold winds and may have to contend with drifting snow. Watch for steep roads approaching the site; they can be treacherous in heavy rains or in icy weather. The dangers inherent in hillside sites, especially in certain parts of the country, were discussed earlier in this chapter. Buy or build on a hillside *only* if you know what you are doing. If you have observed all the precautions, by all means buy a house on a hill if you will enjoy the view. If the climate is mild, your hillside location will be a constant source of enjoyment; even in a cold climate you can always climb the hill in your car if you have a good set of studded snow tires.

E.  IS YOUR LAND ON AN ISLAND?

Your site does not have to be a piece of land surrounded by water to be an island. It may be an area surrounded by high-speed superhighways or just wide heavily traveled streets. These highways may be as much as one-half mile apart or they may be separated by one city block, but in either case you or your children may be literally cut off from any facility you could walk to just as effectively as you would be if you were on an island. If schools, libraries, shopping, and other amenities are located on the other side of the highway from you, then you must be sure there is positive crossing protection for you and your children.

# THE NEIGHBORHOOD AND YOUR SITE

## F. PREVAILING WINDS

This section may seem inconsequential to many people, but a constant breeze which wafts the aroma of the town dump over your backyard is not funny. When you study the zoning map and locate your site, you should also find the direction of swamps, power plants, industrial establishments, or the aforementioned town dump. It is easy enough to find the direction of the prevailing breeze, which is especially important in the summer when you will be living outdoors. If you have a choice of a house or land which lies upwind from any source of air pollution, take it. Several years ago, the town in which the author lives successfully fought a power plant location at the edge of an adjoining town. The plant was to be located at the southwest corner of the author's town, but that is where the prevailing summer breeze comes from. Had the power company not been stopped, the summer sky for several miles downwind would have been covered with a fine haze forever after. Some types of factories are much more polluting than power plants. There are also unpleasant seasonal winds. Try to find a location sheltered from these.

## G. RIPARIAN RIGHTS

If your property is bounded by a body of water, you must be sure of your legal rights and the rights of others regarding access to the water. There may be fishing rights or right of way, or the right of the public to cross the land between high- and low-water marks. Please consult your attorney and be sure that your title to the land is free of encumbrance.

## H. A FINAL THOUGHT ON THE LOCATION

This chapter was meant to be cautionary, not discouraging. If you can find a piece of land or a house which has many things right with it but a few things wrong, by all means buy it. Perfection is hard to find.

## CHECKLISTS

*Your General Impression of the Neighborhood*

1) In the big city renewal area: can you live in the area until it becomes stabilized? This may take years.　　Yes　No

2) Are there suitable schools and green areas nearby?　　Yes　No

3) Do you have to walk through bad areas to get to transportation? Is it safe for your wife to walk to shopping areas and for the children to walk to school? Yes　No

4) A house or land in the suburbs: do you like the feeling of the neighborhood? Are the houses well kept? Do the people look friendly?　　Yes　No

5) The country house: do you like being isolated? Can you see another house or hear traffic from your site?　　Yes　No

*The Quality of the Site*

1) Transportation: See checklists in Chapters 1 and 2. Can members of the family walk to public transportation? In all weather?　　Yes　No

2) Are you prepared to obtain and run two cars?　　Yes　No

3) Are shopping and schools convenient from your particular location?　　Yes　No

# THE NEIGHBORHOOD AND YOUR SITE

4) Utilities: are public water and sewage disposal a requirement for you? Yes No

5) Are the roads maintained? Yes No

*Topography*

1) Sun direction: if you have a choice, in which rooms would you like the morning sun? the afternoon sun? Where would you like your porch or terrace? Are you a lover of houseplants that require direct sun? Yes No

2) Are there any rocky outcrops on the land? Yes No

3) Is there a place on the lot where you can build your house so that it is at least level with the land around it? Yes No

4) Are there any marshy patches or wet ground? Yes No

5) Is your proposed site on a steep hillside? Yes No

6) Do trees overhang the house? If so, can you build your house away from the trees, and if not, will you have some cut down? Yes No

7) Can you make a wide clearing around the house? Yes No

8) Do you like the idea of living on a hill? Yes No

9) Do you prefer a sheltered location? Yes No

10) Do you mind driving on steep roads? Yes No

# CHAPTER 4

# CHOOSING THE HOUSE THAT IS BEST FOR YOU

1. Before You Start the Search for Your House
2. Consider the Salability and Rentability
3. The Choices That Are Usually Available
   A. The New House
   B. Townhouses
   C. Speculative Houses
   D. The Condominium
   E. Cluster Housing
   F. Buying an Older House
   G. Cooperative Housing
   H. Mobile Houses
      i. Mobile House Construction
      ii. Mobile House Parks
   I. Modular Houses

## 1. BEFORE YOU START THE SEARCH FOR YOUR HOUSE

At this point it is necessary to stop and consider your future plans and prospects. The purchase of a first house is an important step forward. It involves great expenditure of money as well as a pledge to a lending institution that you will pay back their loan; that you will pay your taxes and insurance; and that you will maintain the property. It is an important step for former house owners, too, for they must pledge cash and credit.

Therefore, you must consider what the house you intend to purchase will mean to you. If your children have grown and left home, you must decide whether you wish a house large enough for them to visit (and bring the grandchildren); or whether you wish a smaller house with little ground and minimum maintenance. You should determine whether you will be happier living among people of all ages (children included) and sharing their experiences, or whether you would rather "get away from it all" and live among people your own age.

If you are a young couple, with or without children, you should consider potential expansion space. (You may find Section 3 of Chapter 1, "How Large a House Do You Want?" helpful here.) If this is your first house you may be looking forward to moving upward in price and neighborhood. If you have been transferred by your company, it is possible that you may be transferred again. In other words, is this the house you intend to keep?

## 2. CONSIDER THE SALABILITY AND RENTABILITY

Your house should not be considered the same type of investment as a stock or bond. Normally, you will not purchase it as a way of earning an income or making a profit. Yet it is a large investment, and for many people it is the bulk of their entire estate. This means that no matter what your plans are for your house, it must be readily salable or rentable. Events can occur which may force you suddenly to sell or rent. For this reason you are cautioned not to buy or build an odd-looking house that may have a great appeal to you but not to others. Within sensible architectural limits, you should be able to find the house you want. It does not have to appeal to the mass market, but it should be reasonably convenient and conventional. Glamor and great individuality are not enough. The author knows a completely charming house built into a hillside. The kitchen, dining room, and a porch are in the back, which faces a lovely meadow and brook. The front is one floor up and contains the living room and bedrooms. The ground floor, which faces the back, is almost always damp and very difficult to heat in the winter. Many people object to having a kitchen in the basement and everyone objects to dampness and raw chill. It is a difficult house to rent or sell.

There are fads in houses and, like most styles, they fade and tend to date a house. At one time everyone was building a split-level house even on perfectly level lots where they don't belong ; every house had a large picture window, which looked out on nothing in particular and which presented problems in decorating as well as heating. There are many house styles which have lasted a long time and are still good-looking and will be in the foreseeable future. Such houses are convenient to live in, easy to maintain, and will always have a market for rental or sale.

## 3. THE CHOICES THAT ARE USUALLY AVAILABLE

Now you have chosen the localities where you feel you can be happy and can afford to live; and if you have considered the questions asked in Section 1 of this chapter, your next decision is the architecture and general physical appearance of the house you would like to buy or build. You may have a predilection for a particular kind of house but you should at least look at what is available in the entire range of houses for sale. You have an even wider choice if you plan to build. The limiting factor in any case is what you can afford as a first cost and for carrying charges.

### A. THE NEW HOUSE

New houses are available in developments. Most developments contain many new houses as well as others which have been built within a relatively short time of each other, so that the entire neighborhood is more or less of the same age. There is generally not a large price differential between houses in a development, which insures that the neighborhood will be economically homogeneous. Because development houses are often built by an assembly-line method, they usually are less expensive than "made-to-order" homes of equal size.

### B. TOWNHOUSES

Recently, the old-fashioned row (attached) house has made quite a comeback. Development builders are building these "townhouses" in the outer areas of cities and in many smaller communities. Each house has a distinct architectural appearance of its own, which may be Georgian or have a gambrel roof. The outer walls may be brick or some other stone. The entrances are all different. These houses are two or three stories high and many are very attractively laid out. The prices vary of course, but they can be quite reasonable. If you like community living it is suggested that you look at such housing.

# CHOOSING THE HOUSE

## C. SPECULATIVE HOUSES

Many building contractors in smaller communities who usually build houses to order also will build speculative houses to sell.[1] These houses are built singly, generally in established neighborhoods. The intelligent contractor builds such a house to fit into its surroundings. Its architectural style and size usually will be similar to those of the existing houses in the area. This house often will be more expensive than its counterpart in a development. To make up for this, however, it will be in an established neighborhood surrounded by existing lawns, gardens and trees, and will have an established neighborhood living pattern. If you purchase such a house you *know* what the area will look like in the predictable future. Such speculatively built houses are available in a wide price range. They may be found through newspaper advertisements, real estate brokers' listings, or simply by driving through a community and seeing a "For Sale" sign on a newly built house or a house under construction. Sometimes it may be purchased without the payment of a brokerage commission by the builder, with the consequent saving to the purchaser.

## D. THE CONDOMINIUM

The condominium is a comparatively new development in the housing field. It may consist of a group of townhouses, which may be two or three stories high and separated only by a party wall.[2] Or it may be two-story garden apartments with a common entrance to four or more households. Again, it may be a multi-story building with many apartments. Each occupant of a condominium owns his own apartment or unit. A condominium

---

[1] A house is "speculative" if it is built with no particular purchaser in mind.

[2] A party wall between connected houses that separates one house from another and is jointly owned by both.

co-ownership grants the same right to each owner as if he were living in his own house. He has the right to use, occupy, mortgage, or sell his unit. Each owner also has an undivided interest in the property, which is owned in common and which serves all the tenants. All condominium owners must share the expense of operation and maintenance in proportion to their share of the total ownership.

Since the condominium has come into being it has been the subject of a great deal of legislation designed to protect the owner. Such legislation, among other things, requires the separate assessment and taxation of each unit along with its common interests such as land, driveways, common gardens, recreational facilities, and any other facilities which are owned by all the co-owners. It also protects the owner of a unit against any mechanics' lien which may be levied against the property owned in common.

In 1961 the Congress enacted a Housing Act under which the Federal Housing Administration (F.H.A.) may now issue mortgage insurance for individually owned units in multifamily structures. The builder of a condominium usually will arrange to have a single lending institution make mortgage loans available to all purchasers of units, subject, of course, to their personal credit rating. Such loans may be either F.H.A.-insured or conventional; it usually is advisable for the condominium unit purchaser to use the committed lender, although he does not have to do so.

Purchasing a condominium unit appeals to many people. You have the advantages of owning your own house while having your garbage collected, your lawn mowed, your heating plant serviced, your halls and entrances and driveways cleaned and, (where necessary) free of snow and ice. Of course you pay for this service, but only your share of the cost. Because they are built in quantity, condominium units tend to be less expensive than houses of the same size. A condominium involves a type of communal living, where only the inside of your own unit is your own and all other facilities must be shared. It is not for those who want complete privacy in their own backyard.

# CHOOSING THE HOUSE 41

If you are worried about the proximity of undesirable neighbors, there are articles providing for this which all condominium purchasers must sign and observe. All owners are bound by certain rules and regulations which have been promulgated for the common good. A condominium owner who breaks the rules by the flagrant misuse of his premises cannot be dispossessed but he can be brought to court by his fellow owners and compelled by court order to obey the rules or be subject to certain penalties.

If you are interested in the purchase of a condominium you should be familiar with certain basic facts about the project and should have at least the following information before you make any commitments:

1) A complete description of the land showing its boundaries, and a description of the buildings to be built on it.

2) A complete description of how many family units will be involved; how they are to be laid out; apartment or unit plans showing room sizes; how the units are related to each other, and how the various buildings are related to each other.

3) A complete description and layout of all the facilities to be used in common. Will you have a private terrace or garden? Will there be recreational facilities? A pool or tennis courts, etc.?

4) The financial picture. You as a prospective owner must have a very clear and complete picture of what you are getting yourself into after you pay the first cost. The builder or developer will have to give you a certified statement of operating costs prepared by a real estate broker of reputation who is familiar with the management and operation of multifamily properties. This statement should list all anticipated operating and maintenance costs including payroll, public lighting and heating, maintenance of grounds and buildings. You should also know how much your unit will be assessed and the tax rate. You also must be told how and by whom the entire complex will be operated.

As a final caution, please look into the reputation of the

condominium builder or developer—just as you would for any new house that you might buy.

### E. CLUSTER HOUSING

In the past several years many communities with strictly enforced zoning codes have begun to consider cluster housing. Simply described, this means that a developer can build only as many houses on a tract as the zoning code allows but can arrange them in any pattern that he wishes. If he owns ten acres in a half-acre zone, he can build only twenty houses *but each house need not be on its own half acre.* He can arrange his houses so that each individual owns only one quarter acre (which adds up to five acres) and the remaining five acres are owned in common by the twenty families. Such a five-acre plot can be used for many recreational purposes that would not be possible if the tract were evenly divided into half-acre building lots.

Such a development can be badly abused, however, by sloppy neighbors. That five acres of meadow or fields is yours, and yet it is not. Think about your mode of living before you consider getting into cluster housing.

### F. BUYING AN OLDER HOUSE

Perhaps the title of this section should be "Buying a House That Has Been Lived In." There are houses for sale that have been lived in for a very short time and can be considered new. You may not like the colors or planting and may have to add the price of doing these over to your cost, but on the other hand you can see how the house has stood up and you can look for such things as a wet cellar, signs of roof leakage, or bad plaster. You also can obtain the present tax assessment and the cost for heating or cooling the house. Look at the general maintenance of the house and grounds. Lack of maintenance means

FIGURE 1

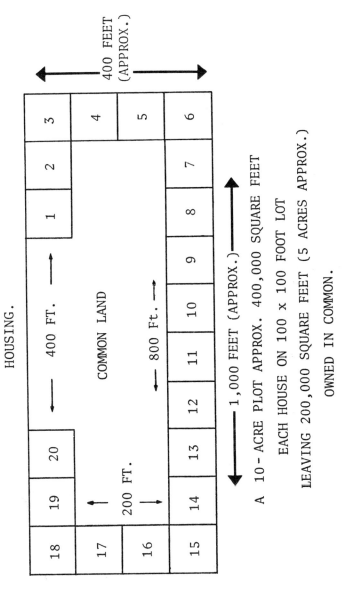

the house has been abused. Do you want it?

After the almost new "used house" comes the house that may be from ten to twenty-five, or even thirty, years old. There are many excellent houses for sale in this category, most of them in established neighborhoods. You may want to "do over" some of the utilities, such as bath or kitchen, but in general a house of this age can be lived in quite comfortably without any major repair or alteration.

Incidentally, houses that were built twenty-five years ago may be better built than they are today. If you are interested in such a house, be sure that its architecture is not dated. It should be of reasonably contemporary style and should not be readily identifiable with any particular period.

Last in choice comes the really old house. This may be an old brownstone in the middle of a planned city redevelopment or it may be a Victorian house in an older but still excellent part of a suburb or it may be an old farmhouse. Look and be charmed but *beware!* Not many people have the talent, the stamina, the patience, or the funds to perform the amount of alteration work necessary to convert a really old house into a comfortable, modern home. Such houses often have bad plumbing or poor wiring; the supporting timbers may be rotted and on a cold day you may have the wind whistling through cracks around the doors and windows and even up through the floor. If you are prepared to perform a good part of the alteration work yourself, you must determine whether you have the skill to do it, whether you are willing to give up all your leisure time and be willing to "camp out" during the time the work is being done. If you think you can afford to have most of the work done by a contractor, obtain an estimate of the cost and *then double it.* These warnings may sound excessive, but be assured that they are not.

G.  COOPERATIVE HOUSING

In a condominium you own your own unit and can lease, sell,

or mortgage it as you wish. In a cooperative ownership you become a member of the cooperative corporation and own stock in it to the extent of your interest. You also own a proprietary lease on your apartment. The cooperative corporation arranges the mortgage financing and employs a managing agent to operate the project and to arrange for the collection of the monthly charges necessary to meet payrolls, maintenance and improvement costs, the mortgage and tax payments.

Cooperative ownership is subject to the general interest of the corporate ownership. You cannot buy your apartment or sell it or sublease it without corporate consent. A cooperative housing project generally is a multifamily management-type structure which may consist of garden apartments or high-rise buildings.

For the past several years many cities and states as well as the F.H.A. and the Veterans' Administration (V.A.) have been encouraging cooperative ownership for lower-income families. Low-interest loans and subsidies and high loan ratios have been provided and many large projects have been constructed.

If you want a housing unit of your own in which you have a financial interest but it is subject to corporate control, and your surroundings are the same as in apartment-house living, you may want to look into cooperative projects. The price per living unit is less than for comparable detached houses or condominiums, and the financing—at least in the less expensive cooperatives—is advantageous. Medium-priced cooperatives usually are found on the outer fringes of large urban centers.

## H. MOBILE HOUSES

The mobile house is an all-year permanent house that is structurally, electrically, and mechanically built to strict standards and is made to be transported over the highways (on its own wheels and chassis) to a building lot where it can be placed on a foundation and connected to the available utilities.

The mobile house must not be confused with the house trailer which all of us see being pulled on the highways by private cars. The trailer, which is much smaller, can be parked in a trailer park and need not be permanently connected to utilities. It is used most often as a vacation home.

The mobile house can be the answer to many people's need for housing. In the past several years construction labor costs have gone up faster and are now higher than any other segment of the labor market and the cost of construction material has not lagged far behind. The result is that home ownership as we know it—the single family detached site-built house—has priced itself out of the market for many people. Young couples who want a house of their own (and don't mind the lack of a second story or basement), retired couples who no longer want a large house, and all middle-income house hunters should look at mobile houses and the modular houses, which will be described later in this chapter.

The Department of Housing and Urban Development (HUD) of the Federal Government has played a leading role in the promotion, the sales, and the popular acceptance of the mobile and modular house. HUD has set as one of its primary objectives the revision of building and zoning codes which restrict the use of such housing. It is also possible to obtain an F.H.A.- or a V.A.-insured mortgage on a mobile house if it is placed on a permanent foundation on land owned by the purchaser of the house. The F.H.A., urged by HUD, has now agreed to insure mortgages to developers of mobile home parks for up to 90 percent of the cost and for a term of forty years. It is also possible to obtain private mortgage financing as well as the installment loans which have been the traditional way of financing mobile houses.

The secret of the ever increasing popularity of the mobile house is its price as well as its convenience. Because it can be moved over the highways to its permanent location, it can be built in a factory by assembly-line methods by shop workers who have the use of lifting devices, power tools, and assembly jigs. These are either not available for the site-built house or are

CHOOSING THE HOUSE 47

not allowed by restrictive construction labor rules. Such assembly-line methods allow you to purchase a mobile house twelve feet wide by sixty-two feet long, containing a living room, a fully equipped kitchen, at least one bathroom, and two bedrooms, for a moderate price. This includes all of your basic furniture as well as rugs and draperies.

1) Mobile House Construction

To overcome some of the original difficulties and the possibility of shoddy workmanship, the Mobile Homes Manufacturers Association has adopted standards of construction promulgated by an American National Standards Committee. These standards provide for the required structural strength and rigidity of the frame to withstand the hazards of windstorm, weather, corrosion, and fire. The standards also apply to the plumbing, heating, and electrical work, which in many cases must also meet the codes of the communities where the houses are to be located as well as the standards of the F.H.A. and the V.A. As for appearance, many of the new mobile houses are now built with peaked roofs, aluminum siding that looks like clapboard, porches, and carports. When the house is set on a foundation and the underside is enclosed by a skirting wall, the last appearance of the old-time "shiny tin box" is gone and it compares favorably with any medium priced site-built house—and in many cases cannot be distinguished from one!

2) Mobile House Parks

Urged on by HUD and other government agencies in response to the acute housing need, many communities are relaxing their building and zoning code restrictions against mobile houses. The western states have long since allowed mobile houses in their residential areas, but the East (especially the Northeast) still retains many of its restrictions. Until recently, many communities in these regions either banned mobile houses completely or relegated them to commercial areas on the "wrong side of the tracks." This is now being changed. Of course, no community will allow a mobile house to be set down in the midst of an area of $50,000 ranch houses or colonial homes, but it will allow for a developer to set off an area in a residential zone and sell lots there for such houses. Many of these mobile house parks are very well planned and executed. There are utilities in place, paved streets, trees, green lawns, and often community recreational facilities such as lawn bowling, tennis, and swimming.

I. MODULAR HOUSES

The modular house, like the mobile house, is factory assembled. It is transported as a unit to its final location and is then set in place on a permanent foundation not unlike that of a site-built house. Modular house single units can be site-assembled as row houses or garden apartments. They can be expanded by adding bedroom units, bathroom units, and so forth. As with the mobile house, the modular house gives the purchaser much more for his money than the site-built house. It can be assembled in interesting architectural patterns and it can be financed conventionally and is eligible to be insured by the F.H.A. or the V.A.

This chapter has endeavored to give an overall picture of the types of houses that you might purchase. Take time to examine those in your price range before you reach a decision.

CHAPTER 5

# THE DEVELOP- MENT HOUSE

1. A Definition
2. The Advantages and Disadvantages of a Development House
3. The Developer-Builder
4. The Physical Structure
5. Inspecting a House During Construction
   A. The Foundations
   B. The Framing
6. A Typical Building Code
7. The Interior Finish
8. The Price to Pay
9. Future Maintenance

1. A DEFINITION

The development house referred to here is one located in a new subdivision, housing estate, manor, or whatever the developer-builder wants to call it. It may be a new house in a newly created neighborhood that was formerly farmland or a wooded area, or it may be located on a lake or other shore area. The developer-builder purchases a large tract of land and subdivides it into building lots in accordance with the zoning regulations of the area. (*You, as a prospective purchaser, must be certain that there are zoning regulations or their equivalent.* Some rural areas don't have any.) These regulations usually specify that the developer install paved roads of certain quality and width, storm drainage, sanitary systems, electric pole lines or underground electrical ducts, and even street trees. They also stipulate minimum lot sizes.

The developer-builder then builds speculative houses on this land and sells them by advertising in newspapers, by mailed prospectuses, by radio, and through other means. The real estate section of a city newspaper (especially on Sundays) will contain many advertisements of such development houses for sale.

2. THE ADVANTAGES AND DISADVANTAGES OF A DEVELOPMENT HOUSE

In previous chapters references have been made to houses in newly developed areas. Such areas need not be outside a city; in fact there are many cities in which recently built subsections and housing developments have created almost a city within a city. The development of some cities can be traced by driving through residential areas a mile or two from the center. As the old houses are left behind, one finds newer houses, some of

# THE DEVELOPMENT HOUSE

which date back thirty years or more. By continuing toward the outskirts of the city proper one reaches newly built homes and even vacant tracts with development just starting.

There can be two kinds of houses in these areas. The first is the individual house or group of two or three being built on a road or street that is a part of the city's existing street plan. A builder may be constructing these on speculation or for individual purchasers. The other kind is the house with which this chapter deals. These are the rows of fifteen, twenty, or even hundreds of houses that have been recently built or are being built on new roads, which have been made by the developer. These roads can form a right-angled gridiron pattern or can be curved to follow land contours. If these roads meet the municipality's requirements they are taken over and maintained by it. The houses may be of one design with the only variation being the decoration on the garage door, the color of the shingles, or the location and pattern of the front door. There are also developments offering different styles of houses.

If you are considering a house in such a new section you must be aware of many things:

(a) You will become a member of a new community where everyone else is new. There will be no established patterns of neighborhood or social activities. This may be a good thing, for it can give you an opportunity to help organize the neighborhood.

(b) If the subdivision is large enough, the municipality or the governmental body under whose jurisdiction it falls may treat it as a separate tax rate area. This is because the local government may be put to great expense to integrate the new section into its sanitary sewer system, its storm drainage system, and its town road system.

(c) These new areas, if they are large enough, require new schools which must be built by taxpayers' money.

(d) Public transportation may take some time to catch up with the new population growth.

(e) Conveniences such as shopping may lag behind unless

the developer includes a shopping center or a small local merchandise center in his plan.

(f) If the development is far from a town center, the road network connecting it to the center may not be complete or even planned yet.

Decide which of the foregoing considerations refer to your location and which are important to you. You must keep in mind that many new developments or housing estates are noted for their spirit of freshness and energy and neighborliness. All the people are new and most are in the same economic and social bracket. In the less expensive areas many of the people will be younger couples with small children. They can produce built-in baby sitting and nursery school service and many other amenities which are not readily available in older neighborhoods.

## 3. THE DEVELOPER-BUILDER

It is important for you to acquire some knowledge of the reputation and financial status of the developer-builder. This information may be obtained from the mortgage man at the local bank, the local building inspector, or the real estate broker. If the mortgage is an F.H.A.- or V.A.-insured one, the nearest offices of these agencies will know about the builder. You even can obtain a credit report concerning his financial status and reputation. After all, the bank or the builder will get a credit report on *you*.

You may ask why you need a credit report or other evidence of the builder-developer's financial status and reputation. The answer is simple. There is a long history of unethical developers who "walked out" after a house or a number of houses were completed and sold. They may leave the houses with bad foundations, defective drainage systems, leaky roofs or other defects, or they may leave roads that break up during the first cold weather or after heavy rains, and the purchasers have nowhere to turn for redress.

You should also be aware that a builder who is in shaky financial condition may not be able to pay his bills for labor and material. If you buy the house you must be sure that your attorney has carefully investigated to insure that all bills have been paid; otherwise you may find your house subject to a mechanic's lien.[1] However, certain waiting periods are stipulated in the various state laws, after which a mechanic's lien cannot be filed.

Reputable developer-builders will guarantee a new house against poor or defective workmanship and material for a certain fixed time up to a year. If you get such a guarantee, read it carefully to see that it actually promises that the builder will come back to repair any defective work *at no cost to you.*

Find out what developments or houses have been done by the builder in previous years. Drive through these developments if they are near enough. Note whether the roads are still in good condition after a few years. Look at the state of the houses and lawns and shrubbery. Perhaps the real estate broker will know someone who lives in a house in a previously built development. Many of the large developers have sales organizations of their own whose entire mission in life is to sell you a house. Don't be stampeded. Ask all these questions. Don't forget that you may be spending a lifetime of savings on this house. You must be a well-informed buyer.

## 4. THE PHYSICAL STRUCTURE

We now come to the actual physical structure of the house.

---

[1] All states have laws which protect workmen and material suppliers against unpaid wages or bills. Such a person can file a lien against a house, which gives him an interest in it which is prior to your own.

Builders of large housing developments very often build their houses by an assembly-line method. One crew digs for the foundations; another crew places the concrete or block foundation walls; there is a crew of framing carpenters and roofing carpenters and so on. This is fine and economical for the builder and very often good for you, too, if the workers know what they are doing. It sometimes happens, however, that some mass builders subcontract their work on a piecework basis which puts a premium on the speed of production. In this case the buyer should investigate the quality of the construction he is getting.

## 5. INSPECTING A HOUSE DURING CONSTRUCTION

If you intend to buy in a development which is still under construction, it is an excellent idea for you to carefully examine houses that are in various stages of completion. Some of the parts of the construction which you can look into are the foundations and the framing.

### A. THE FOUNDATIONS

Many development houses, especially in milder climates, do not have any basements; the house is built on a flat concrete slab which is poured in place on the ground. Such concrete slabs usually have the heating pipes buried in them. If you can see this foundation as it is being built, see if the ground looks dry and solid. The bare ground should be covered with a layer of crushed stone to provide drainage. The crushed stone should be covered completely with a sheet of plastic to prevent ground moisture from seeping into the concrete. The heating coils and (most important) a reinforcing steel wire mesh must be in place before the actual concrete is placed. The wire mesh should be of heavy steel wire to form squares usually 6 × 6 inches. It

# THE DEVELOPMENT HOUSE

must be kept off the bottom of the slab by some kind of support and the concrete slab itself should be at least six inches thick. In colder climates the entire floor slab should be surrounded by a perimeter wall that extends into the ground below the frost line. (See Figure 2.)

The same cautions regarding foundations must be observed in houses with basements. Foundation footings should be placed under the foundation or basement walls and the bottoms of these footings should be on good solid earth. A foundation built on filled land or swampy land or moist clay can sink or slip, unless special foundation precautions (such as piling or spread footings) are taken. Most localities have building inspectors who will check this, but your own personal observation of how the builder observes the rules of good construction will be reassuring to you. (See Figure 3.)

B. THE FRAMING

The framing is the basic skeleton structure of your house. If you can get to see a house under construction, look for several things. The studs or upright supports should be of 2 × 4s which usually now measure about 1¾ inches by 3½ inches. They should be not more than 16 inches apart and should stand on a 2 × 4 "sole" which is laid on the concrete slab and fastened to it—or nailed to a rough flooring if the house has a basement. There should be double 2 × 4 top "plates" over the studs; these will support the roof rafters and the ceiling beams. Make certain everything is being securely nailed. Inspect the lumber to see whether it looks reasonably smooth and free of knotholes and other defects. Try to see whether the roof rafters are properly fitted and securely nailed to the top of the stud wall. (See Figure 4.)

In the case of a brick or masonry block house, the walls should be at least 8 inches thick. It is better to have a space between the exterior and interior masonry to prevent seepage of moisture through the wall. Check for solid mortar joints and

FIGURE 2

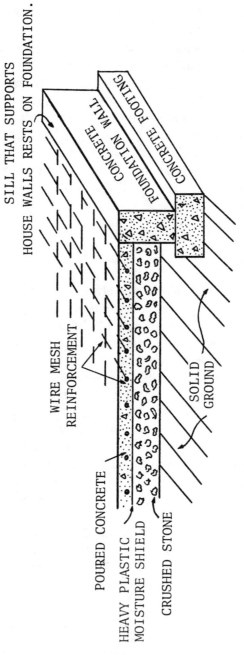

SECTION OF FOUNDATION FOR FLAT CONCRETE SLAB

FIGURE 3

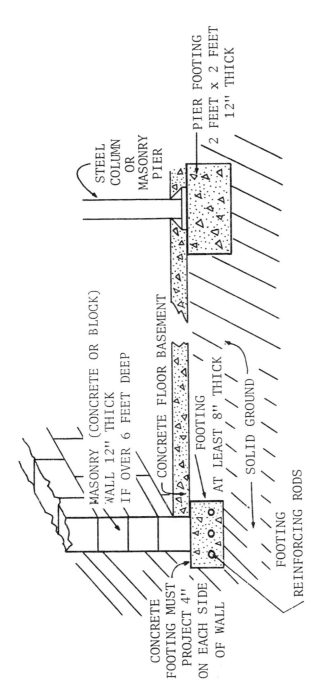

SECTION OF FOUNDATION WALLS & FOOTINGS

FIGURE 4

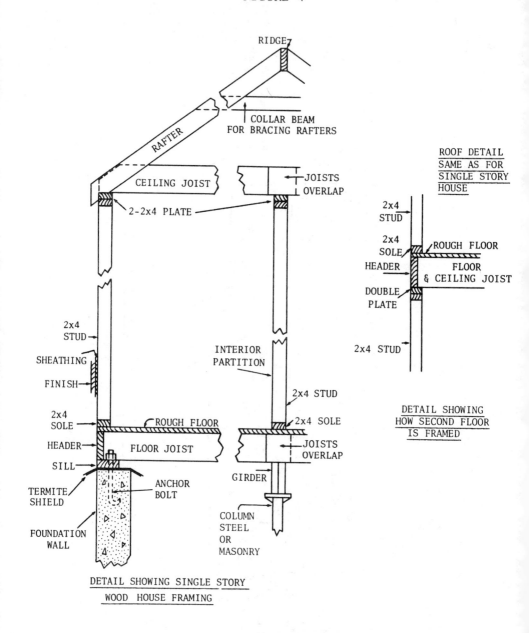

A STRUCTURAL FRAME
FOR A TYPICAL WOOD HOUSE

proper ties between the inside and outside portions of the wall. It is advisable that the interior walls of all brick or masonry homes be "furred," or separated from the outer masonry wall by strips of wood or metal that will form an air space between them. If this is not done there is every chance that you will have damp, cold walls.

Look at the roof and the gutters and leader pipes. A good material for a roof is heavy asphalt shingle. It is long-lasting, fairly fireproof, and if well laid over a plywood or other solid base it will keep your house dry for fifteen or twenty years. The gutters and leader pipes in development houses are usually of aluminum. Copper is much better but also much more expensive. Be sure that the metal is not painted galvanized iron, which won't last more than a few years. Many houses have wood gutters which are satisfactory if they are so built that an overflow will not seep into the outer wall. (See Figure 5.)

## 6. A TYPICAL BUILDING CODE

An outline of a typical small-town building code will provide you with the most important points of acceptable construction (these are not exact quotations):

(a) Wall footings of poured concrete should be not less than 16 inches wide by 8 inches thick, or in the case of a one-story house they should be not less than 12 inches wide by 6 inches thick. They should be placed on solid undisturbed ground.

(b) Foundation walls of concrete should be not less than 8 inches thick and walls built of masonry units should be not less than 8 inches thick if only 6 feet high and not less than 12 inches thick if more than six feet high. The top four inches of hollow masonry walls shall be filled in solidly with cement mortar. Masonry bearing walls shall not be less than 8 inches thick for a one- or two-story house.

FIGURE 5

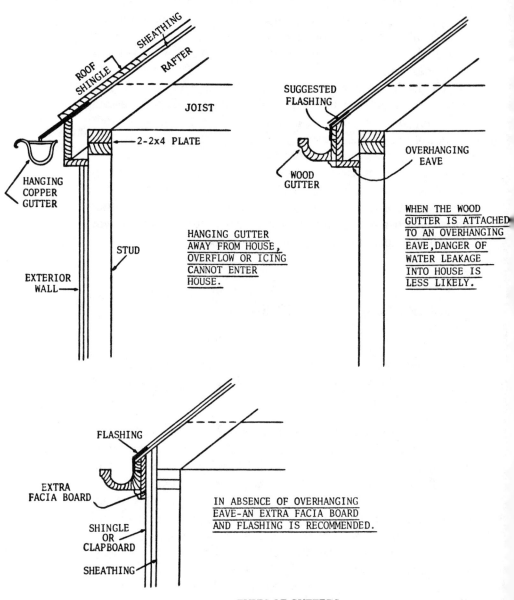

TYPES OF GUTTERS

(c) Studs or vertical framing members shall be not less than 2 × 4s, not more than 16 inches apart, and should have double 2 × 4 plates at top. Floor joists shall be of the proper sizes and shall be properly spaced in accordance with the load they will support and the clear space they will span.[2] (In the ordinary development house the sizes will be the minimum allowed. For instance, 2 × 8 floor beams spaced 12 inches apart can clear a maximum span of 12 feet. If this size beam is used on a longer span the floor will be shaky. A span of 15 feet will require 2 × 10 beams spaced 12 inches apart.)

(d) Roof rafters shall be vertically supported at the ridge or shall be tied together with collar beams of 1 × 6 inches, spaced not more than 5 feet apart. (See Figure 5.)

(e) Framing over openings such as windows or doors shall be as follows: Spans up to 4 feet—two 2 × 4s; spans up to 6 feet—two 2 × 6s; and so on.

(f) Exterior walls of frame buildings shall be sheathed with at least 1 inch thick wood or 1/2 inch thick gypsum board or fiber board, and over this there shall be at least 1/2 inch thick wood siding or 3/8 inch wood shingles or 5/32 inch asbestos shingle.

There are also code requirements for roofs, furnace rooms, plumbing, electrical wiring, and other parts of a house. Unless one is familiar with these trades it is difficult for the layman to check on whether the work is being installed correctly. The prospective purchaser must depend on the local building inspector and the fact that electrical and plumbing codes are probably more strictly enforced than codes relating to any other part of building construction.

---

[2] None of this material is very technical, and a study of the simple sketches will show you exactly what is meant.

## 7. THE INTERIOR FINISH

If you cannot inspect the actual construction or if it seems too complicated for you, then the next best thing is to examine the appearance and quality of the inside finished work. Inspect the woodwork to see if the joints fit tightly. Look at the door knobs and door hinges to see if they are properly fitted and give an appearance of solidity. Open a few doors to see whether they stay where you leave them or move. If they move they are improperly hung. Open and close the windows to see if they slide evenly without binding. Look at the thickness of the front door. It should be at least 1½ inches and preferably 1¾ inches. The exterior hardware should be brass or stainless steel. Close several doors and speak loudly in one room to determine whether the walls are reasonably soundproof. Get the feel of the general solidity of the house.

## 8. THE PRICE TO PAY

It is difficult at this time of wildly rising prices to state boldly that a house of such-and-such dimensions and so many rooms should cost a certain amount. What is more, prices vary widely over the country and depend on construction union practices (or non-union labor), rigidity of zoning and building codes, the distance from the source of the materials of construction, and other factors. Nevertheless, some comparisons may be made. A careful study of the real estate section of any large Sunday newspaper will give you a fairly good idea of the houses that are available within your price range.

This wide variation in prices in the same metropolitan area shows that you must carefully evaluate what you are buying. Unless there is a compelling reason to live in a particular location you should do some comparison shopping. If a prudent

# THE DEVELOPMENT HOUSE

housewife goes to one shop because it has fresher vegetables or sells bread, butter, and eggs for less, then certainly you must do the same when buying as important a commodity as a house. The checklist at the end of this chapter may help you.

## 9. FUTURE MAINTENANCE

One other important consideration in purchasing or building a house is the cost and trouble of its future maintenance. Copper water lines are a necessity where there is "soft" water, which corrodes all piping except copper or red brass. The furnace should be oversized so that it does not overwork and so that you can add a room to the house or finish an attic and heat it. Find out the capacity of the electrical feeder line. Many communities insist on a capacity of 100 amperes, especially if there is an electric stove. Be sure the gutter and leader pipes are of noncorrosive metal. If you are in a cold climate, see if the windows and doors are weatherstripped. Your common sense will tell you what is likely to cause trouble.

(*see* Chapter 19 for greater detail on choice of materials.)

## CHECKLISTS

*Evaluation of a New Community*

1) Do you want to live in a neighborhood where everybody is new?   Yes   No

2) Have you investigated the present and *future* tax rate?   Yes   No

3) Are you likely to be assessed for new sidewalks, sewers, or other municipal improvements? Yes No

4) Is adequate public transportation available? Yes No

5) If not, do you have or want a second car? Yes No

6) Is local shopping adequate for normal needs? Yes No

7) Are schools or bussing to schools available? Yes No

*The Price to Pay—Comparison Shopping*

1) Are you willing to pay more for a comparable house in an exclusive neighborhood (which may be "exclusive" only in the developer's mind)? Yes No

2) The Physical Comparison (to be completed for each house that you are considering):
Size of the land. Is it large enough for a garden and a children's play area?
Is the architecture and general appearance pleasing?
Is there any gadgetry, such as a conversation pit or a cathedral ceiling?
Do you want them? These styles may be eyecatching for the moment only.
Number of bedrooms and average size.
Number of full baths.
Number of half baths (basin and water closet).
Size of living room.
Size of family room (if any).
Size of dining room (if any).
Size of kitchen. (Is it a combination

kitchen and dining room?)
Is there a basement under the entire house?
Size of the garage. (Is there room for tools, bicycles, mowers, etc., besides space for cars?)
Kitchen equipment.
Laundry equipment.
Extras: storm windows, screens, carpeting, landscaping, sauna, others.
What is the area of the house in square feet (excluding the garage and the basement)?
How close are you to your neighbor? Do you have enough privacy?

# CHAPTER 6

# THE SPECULATIVE HOUSE BUILT BY A LOCAL BUILDER

1. What It Is
2. How to Find It
3. How to Buy It
4. What to Look for in a New House
   A. The Exterior
   B. The Interior
5. The Zoning

## 1. WHAT IT IS

Most towns have local builders who build one or more single family houses for sale as part of their annual business. Such a builder can do good work and is accustomed to dealing with architects and homeowners on an individual basis. He usually owns a few building lots in the area and between contract jobs he will start to build a speculative house to keep his crew, usually small, busy. This house generally will be built in or near an established neighborhood and usually will reflect the architectural style and size of the houses around it. Many well-established communities with strict zoning and building codes are out-of-bounds for developers of large tracts. These communities do not encourage developers of large tracts and usually the land and building costs are uneconomic for them.

Consequently, if you would like to buy a new house in an established community you are likely to end up with a house built on speculation by a local builder.

## 2. HOW TO FIND IT

Very often such houses are individually advertised in an effort on the builder's part to avoid a broker's commission. These advertisements usually can be found in the Sunday real estate section of the nearest large town newspaper and in the local newspaper. After you have chosen the general locality where you wish to live, a study of the local newspapers will be rewarding.

Another method of finding this house is to obtain a map of the community and drive around. Many builders will put "For Sale" signs in front of their houses to avoid a broker's commission by means of a direct sale. Be sure if you purchase directly that you obtain the benefit of the direct sale, which

# THE SPECULATIVE HOUSE 69

is saving the builder 5 percent or more of the sale price.

The easiest way to find a house is through a broker. Many city brokers have branches in outlying suburbs or a working arrangement with local brokers.

## 3. HOW TO BUY IT

You must be sure that the broker you are dealing with has the authority from the owner of the property to transmit offers and acceptances and to accept a deposit. The deposit, seldom more than $1,000, binds the house until the signing of the formal contract. Before you pay any deposit, especially if the house is still under construction, be sure that you know what you are getting and *when* you will obtain beneficial occupancy.[1] Sometimes a broker, because of insufficient experience or lack of familiarity with the final terms of the sale, will not be able to give you the information you require before you make a deposit. This is not too important when you are buying a finished house and you can see what you are getting. However, it is unwise to pay deposit money on an incomplete house, unless there is an understanding that you can have your deposit refunded within a certain specified time before the contract is signed, if further investigation on your part disclose that some basic part of the deposit understanding will not be fulfilled. Chapter 16 explains the terms and conditions of the contract which you will eventually sign and of the closing which will make you the owner.

One further word on "how to buy": You must be sure before you enter into an understanding to purchase a speculative house that the terms and amount of the mortgage money fit your financial requirements. You can make such arrangements

---

[1] This is the date when you are in and the workmen are out, except for small repairs or adjustments.

personally through F.H.A. or V.A. or with a bank or insurance company or a building and loan society. You also can have the builder or broker arrange the mortgage loan.

## 4. WHAT TO LOOK FOR IN A NEW HOUSE

Your adventure in house purchasing will be less dangerous if the house that you are purchasing is completed or near completion. Even if you cannot see the structural frame, you can see how it is being finished.

### A. THE EXTERIOR

Begin by examining the landscaping. Look for planting (foundation or otherwise) and walks. Do you have a suitable driveway? Is there a lawn or is the area surrounding the house graded and prepared for a lawn? Some builders have been known to strip the best topsoil from a building lot and leave only the marginal soil on which it would be difficult to grow a lawn. Perhaps you can dig a trowel or a sharp stick into the ground. Topsoil should be at least six inches deep.

Examine the exterior of the house. Look at the tops of the window and door frames to see if there is a metal flashing over them. (*see* Figure 6.) This flashing forms a protective barrier against water coming into any crack between the top of the window frames and the vertical wall. Examine the gutters and leader pipes to see if they are of noncorrosive metal. If the gutters are wood, check to see that they are attached to a board fastened to the exterior wall and not to the wall itself. (*see* Figure 5.) An overflowing or iced gutter unless it is isolated from the exterior wall can leak down into the wall. Examine any exterior molding to see if it is properly fitted. Look at the paint to see if it seems to cover the wood properly

FIGURE 6

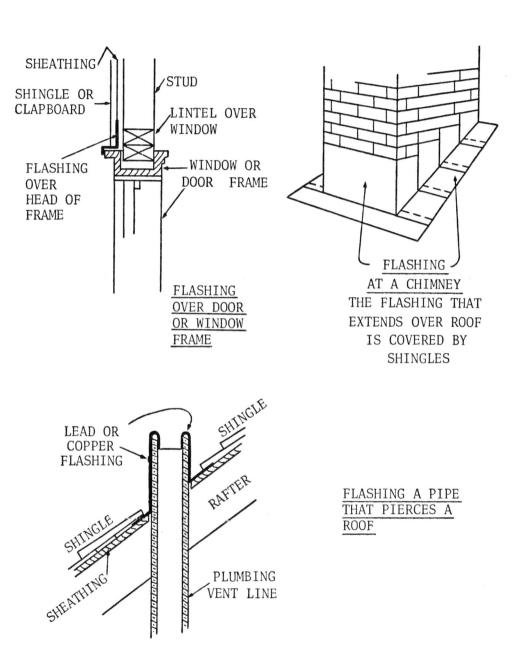

and neatly without drips and spaces. If it is a brick house, look at the mortar joints to see if they look solid and have been rounded or otherwise smoothed. Be sure the interior wall has been "furred" or separated from the masonry outer wall by strips of wood or metal that will form an air space between them—or you may have damp, cold walls.

## B. THE INTERIOR

Chapter 5 and 19 go into some detail on interior finish. In a housing development there is usually a sample house for you to look at, and you can see what will be furnished as kitchen equipment and cabinets and bathroom fixtures and cabinets. If this equipment is not in place when you are looking at a single speculative house rather than a house in a development, you must be sure to inquire as to exactly what you are going to get in the completed house. If you cannot be shown the actual fixtures and appliances, then you should see catalog illustrations of the stove, the refrigerator, the kitchen sink, the laundry equipment, and the bathroom fixtures and cabinets. Your sales contract should refer to these catalog pictures or numbers.

If lighting fixtures are not in place, the usual custom is for the builder to give you a fixture allowance and to send you to a lighting fixture shop where you can make your own choice. If you choose fixtures which exceed this allowance you must pay the difference. Check the locations of the electrical outlets (see Figure 7). If they are not yet in place, you may have a choice of location. Be sure there are enough, especially in your kitchen. How much is enough? Count all the possible electrical appliances or accessories you may ever use in any room, then add two.

It is important to check for soundproofing between living spaces—especially when you are buying a townhouse or a condominium. Sounds caused by a neighbor even in ordinary everyday living can be intensely irritating.

In most individually built speculative houses the builder will allow the purchaser to choose interior finishes. You will probably get an allowance of so much per roll for wallpaper and be allowed almost any color for painting. It may be possible to choose your floor or floor finish if the house is just being completed. Floors vary by regions. The floors may be of various woods, terrazzo, ceramic or asphalt or vinyl tile, and so forth. You must ask a local flooring firm what is best for your area.

The above advice and comment is to help you to ask the right questions and to look for and obtain from the builder all the necessary appliances and finishes that you will require for comfortable living. Be sure before you buy that you know what you are getting!

## 5. THE ZONING

It is important to mention zoning again. Even though you are looking at a house which is being built or has been built in or near an established neighborhood, it is still wise for you to check the zoning regulations regarding your specific area and the general area. Sometimes zoning is upgraded, and the lot on which a speculative house is built may have been purchased before the upgrading. Thus the lot may be smaller than what is called for by the new zoning code. The lot (and the house upon it) becomes an "allowed nonconforming use," to which special regulations regarding additions or improvements may apply. Check with the local building inspector.

Note also the boundary lines of the various zones with relation to the location you are considering. These are not difficult tasks. Zoning maps and regulations are readily available. Brokers and builders will have them. Don't be afraid to ask questions. It's your money!

FIGURE 7

## SUGGESTED ELECTRICAL OUTLETS

## FOR LIVING ROOM:

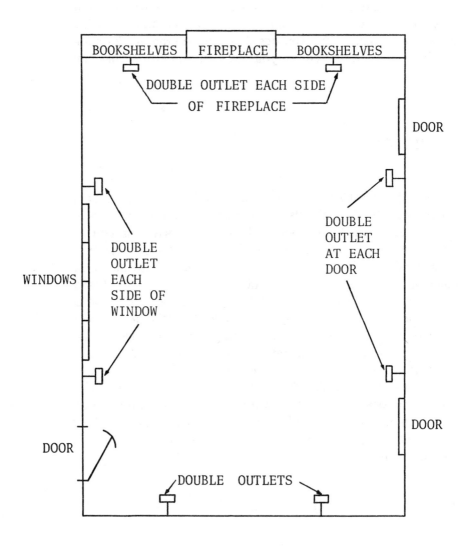

FIGURE 7A

## A CORNER 12 × 12 FOOT BEDROOM:

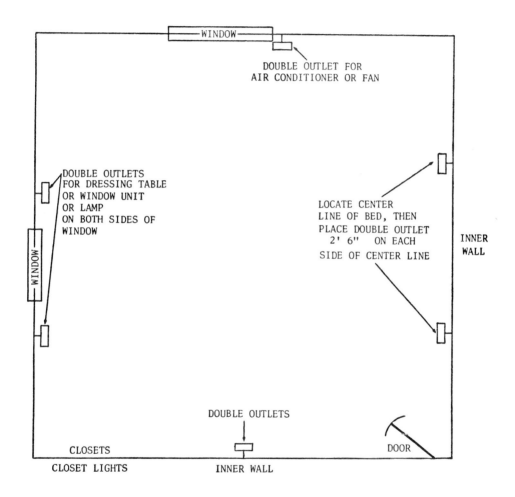

# CHECKLIST

*What to Look for in a New House*

The driveway: gravel, crushed stone, or paved.

Planting: trees and shrubs; foundation planting.

Lawn: look at the topsoil.

Leaders and gutters: look at the wood gutters. Is the metal noncorrosive?

Examine tops of window and door frames for flashing.

Examine the paint. Does it look like a good cover?

Masonry walls: examine the brick or stone joints; make sure interiors of exterior masonry walls are furred.

*The Interior*

Kitchen: either see the actual fixtures or pictures of them; check kitchen cabinet space and counters.

Bathrooms: same as for kitchen. Ceramic tile is the best finish for bathroom floors and wainscoating.

Lighting fixtures: obtain a sufficient allowance for good fixtures.

Electrical outlets: are there enough?

Interior finishes: dark colors should be two coats over an undercoat. Get a sufficient wallpaper allowance.
Make sure there is nonskid finish on your floors. (Also *see* Chapter 19)

## CHAPTER 7

## BUYING A HOUSE THAT HAS BEEN LIVED IN

1. The Available Choices
   - A. The New House under Temporary Lease
   - B. The Almost-New House for Sale by the Owner
   - C. The Older House
2. Choosing the Best House for You
   - A. If You Can Lease before Purchasing
   - B. The Almost-New House
   - C. Examine and Carefully Assess the Older House

## 1. THE AVAILABLE CHOICES

### A. THE NEW HOUSE UNDER TEMPORARY LEASE

Chapters 5 and 6 discussed the buying of brand-new houses under various conditions. Some houses have been lived in for a very short time and may be almost new. There are speculatively built houses which are leased for a time if the builder cannot obtain a purchaser within a reasonable period.[1] These houses are for sale and can be possessed at the end of the lease term. Local brokers will know of such houses and will know the terms of the lease. You may even be able to lease it yourself. You may try to obtain an option to purchase at the end of a fixed term at a certain price; if you can't get this, you may be able to get an option to meet any offer that the owner may obtain. Of course, your lease will protect you to the end of its term, but at that time, if you have no option clause, you may find that the house has been sold from under you.

### B. THE ALMOST-NEW HOUSE FOR SALE BY THE OWNER

In all suburbs of large cities there is constant moving of corporate executives. Most of these people have not yet reached the top echelon and are therefore subject to transfer. Many people of this class buy or build houses, in the hope that they will remain where they are for a number of years. These houses are very often almost new, and some of them may have been lived in by as many as three families within a very few

---

[1] A lease is a legal document which specifies that a certain piece of property is to be rented or occupied for a certain fixed time for a certain fixed amount. Anyone who rents a house is advised to obtain a signed lease.

years. If the owner has sufficient notice of his transfer or is selling the house for other reasons, he may advertise it for direct sale. Recently, many large companies who transfer their executives have made arrangements to take over such houses, if they are not sold within a certain time. They pay the owner the appraised value of the house and then dispose if it—frequently selling it at their cost in order to get rid of it quickly. There are, of course, houses for sale for other reasons, and these, too, are generally listed by all the area brokers. If you are interested, ask realtors about this type of house or look for one in the real estate sections of local and city newspapers.

## C. THE OLDER HOUSE

The older house may be thirty or more years old. At first glance it may look contemporary and it is frequently in a well-established neighborhood. It may come on the market because of retirement or transfer or because it is too large for the present occupant. This house may often be purchased at a reasonable price, but it should be examined carefully in order to determine the repairs and modernization which will be necessary. The cost and inconvenience caused by such work could be more than the initial saving in the purchase price and such a house may be difficult to finance. It is also important that you look carefully at the general appearance of the exterior. If the house was well designed with simple lines and proportions, its age is indeterminate. Some houses, however, date themselves by "temporary" design which was fashionable when they were built but which is no longer so.[2] This can be satisfactory if you like the style, but it may make it difficult if you ever have to consider resale.

---

[2] See Chapter 10 on Architecture regarding such temporary styles.

## 2. CHOOSING THE BEST HOUSE FOR YOU

### A. IF YOU CAN LEASE BEFORE PURCHASING

Sometimes it is possible to lease a house for a year or more with an option to buy it. This can happen not only with a new speculative house but also with an older house when the owner, for various reasons, must either sell his house or lease it if he cannot sell immediately. This situation presents a great opportunity for you to gain the experience of living in the house without having to commit yourself to its purchase. You can then discover for yourself the advantages and disadvantages of the house without depending on hearsay or the optimistic reports of the builder, owner, and/or broker. There are a number of important matters which you should check:

*First:* The quantity of fuel necessary to keep the house comfortable during the cold months or the electricity necessary to keep it cool in the warm months.
*Second:* The general comfort of the house. You should note any drafty spots or rooms which are hard to heat or cool.
*Third:* The direction of the sun. Notice whether your terrace or porch faces west or partly west, which will give you the full benefit of the hot afternoon sun in the summer. Perhaps you prefer this to getting the sun in your living room, but check anyway.
*Fourth:* The general sturdiness of the construction. You can actually see if you have a wet cellar during a wet spring or a leak in the roof during a heavy rainstorm. Look for damp spots in the ceilings or walls over the windows. Check for sound-leakage from room to room; you should not be able to hear more than a murmur of voices when people are speaking in a conversational tone in an adjoining room with the door closed. Can you hear a shoe drop above you? It should sound muffled. Does the floor seem to vibrate when you walk or run across it?
*Fifth:* Whether the layout of the rooms, their size, and their

# BUYING A HOUSE

relationship to one another and the outdoors meets the needs of your mode of living. If the house is physically satisfactory but is inconvenient in some way to the point of daily annoyance, hesitate before committing yourself to purchasing it. If the house meets your needs otherwise, possibly such an annoyance can be remedied. Perhaps the layout of the kitchen bothers you or you can't watch television without waking the baby. These are conditions which may be taken care of by some minor remodeling. (Chapter 12 on Remodeling and Chapter 10 on Room Arrangement for Convenience may be of interest to you).

Having the chance to live in a house before purchasing it is an opportunity you should not overlook.

## B. AN ALMOST-NEW HOUSE

Houses that are almost new come on the market rather frequently in the better suburbs of the larger cities. These houses which have been occupied and sometimes reoccupied by transient executives can present a dismal picture if they have not been carefully maintained.

It is very easy for a young family with small children to neglect a house; and this is especially true if they feel that this may be only a temporary home. A well-built house can take a reasonable amount of abuse, but badly scratched floors, gouged plaster, neglected exterior painting, or poorly maintained shrubbery and lawns are expensive to restore. Try to make an inspection when the lady of the house is not present. Look at the linoleum or other floor covering around the kitchen sink, basins, and water closets to see if it shows signs of rot. Open kitchen cabinets and the refrigerator and stove; even a relatively new stove can be so badly encrusted that you may have to buy a new one. If you get a feeling that the house has been used very hard and that it has been neglected, it may be wise to look further before committing yourself. Outward signs of neglect may be a sign of more serious trouble.

## C. EXAMINE AND CAREFULLY ASSESS THE OLDER HOUSE

If you are considering an older house, there are several basic questions which you must answer. You must make up your mind as to the order of their importance.

*First:* The exterior appearance. Some houses look as if they belong where they are; even an "old-fashioned" look does not change this. Other houses simply look dated. They were built thirty or forty years ago with no particular attention to proper proportion, material, or style—and they look it. The question for you is: can you live with it and be happy or do you want to go through a major face-lifting? Can you purchase the house at a low enough price to warrant the expense of alteration? You must remember that such a job will call for the services of an architect or a good designer and that the house may be barely habitable for a number of months. You must consider one other important question. If the house is architecturally altered to comply with the requirements of good design, will it be badly out of keeping with the surrounding houses and the neighborhood? Overbuilding in a neighborhood is a poor investment.

*Second:* Room size and arrangement. You must determine whether the room sizes and their arrangement meet the demands of contemporary living. Many older houses have small rooms, practically no closets, and were not arranged for the comings and goings of present-day children, for outdoor living, for hobbies, and for extensive entertaining. The house must be convenient.

*Third:* The equipment of the house. The kitchen and bathrooms may be so old-fashioned you will want to modernize them. Be sure to check the cost of doing this very carefully; present-day prices may be prohibitive, so perhaps you can have a partial job done. Make up your mind how little you can have done and still be reasonably satisfied. You may be able to enter into an orderly program whereby you do so much each year.

*Fourth:* Physical structure. The entire physical structure of the house should be carefully inspected. You should look for rotted sills, for evidence of termites, and for cracked walls or ceilings. Turn on the kitchen and bathroom faucets at the same time to see if you have enough pressure. Check to see whether the flow of cold water in a shower slows down abruptly when you turn on a cold water faucet in the basin or kitchen sink. There have been cases of painful scalding when this has happened. If you don't have enough pressure you may be faced with an expensive replacement of water piping. Examine the electric wiring and the main panel board for rusting of BX cable and for sufficient capacity. Do you have enough outlets? Examine the furnace and the fuel burner. How old are they? A mechanical burner that is over twenty years old undoubtedly has lost efficiency and spare parts for it may be out of stock. Look at the sheetmetal stack from the furnace to the chimney. Does it look rusted? A leaky stack is extremely dangerous.

When you have completed your inspection, obtain a reliable estimate on the cost of bringing the house up to a state of reasonable repair and architectural quality. Add this amount to your purchase price; then ask yourself: Is this really the type of house I want, or would I do better to look into a new one?

# CHAPTER 8

# REHABILITATING AN OLD HOUSE

1. Before You Buy — Investigate
2. The Structural Frame
3. Heating, Plumbing, and Electrical Wiring
4. The Woodwork and Floors
5. Plaster
6. The Roof
7. Your Time and Effort
8. The Not-So-Old House

## 1. BEFORE YOU BUY—INVESTIGATE

The houses which will be discussed here are over fifty years old and are more likely to be closer to a hundred years old. The house that has been built within the last fifty years is, comparatively speaking, of modern design and the construction is much the same as it is now (except perhaps better). However, the old house—which may have been partially rehabilitated by a former owner or by several former owners but which is now barely habitable—requires a thorough examination by an expert eye before you can commit yourself to buying it. If you are seriously looking at such houses, the chances are you will buy one no matter what. But at least you should know your problem and your probable future outlay in time and money.

## 2. THE STRUCTURAL FRAME

Taking the important components of a house in order, the first item to be examined is the structural frame. Don't worry about the thick coats of old varnish on the woodwork—you can always remove them; but if the structure itself is in poor condition, be prepared to spend a great deal of money. Such structural repair work can very rarely be done by an amateur and requires professional help. Whether it be an old brownstone or row house in a city, an old farmhouse or just an old house anywhere, the first thing to look for is the condition of the sills, the studs, the girders, the joists, and the rafters. The sills are the wood members that lie flat on the foundation walls and upon which the house rests. (See Figure 4.) Lying as it does on a stone foundation wall, and being subject to constant dampness and possibly termites or carpenter ants, the sill is likely to be either riddled by termites or rotted. To a lesser extent, the same is true of girders and joists. Look for charmingly tilted floors. They spell danger. Look for any wooden portion of the house which is in direct contact with the ground. This is an invitation to termites or ants. Get into the basement or crawl area and examine these bottom timbers closely. Stick a sharp knife into them. This will not hurt them if they are sound but the knife will go in easily if they are rotted or eaten. If you can't get at these timbers because the basement ceiling is plastered or otherwise covered, then walk firmly diagonally across each floor and really put your weight down hard. Do try, though, to find an uncovered part of the basement ceiling. Look for bulges in the outside walls, for a bulge in a brick wall may mean a settling foundation wall. Look for severe cracks in ceilings and walls. Look for a swaybacked roof line. Examine the roof rafters to see if any of them have started to bend or crack.

If you note any of these indications of a weak or rotted structure, you must be prepared to employ professional help such as carpenters and masons.

## 3. HEATING, PLUMBING, AND ELECTRICAL WIRING

Many old houses still have large hot-air furnaces with floor registers or, if they were rehabilitated some time ago, steam boilers and radiators. In the first case be prepared to install a new heating system. An old hot-air furnace with its uneven heating and its dust and dirt simply cannot be lived with in this day and age. A steam system can be endured but *the boiler and the piping must be carefully checked for rusting and leakage.* With the proper tools an amateur can repair heating lines, but it takes a lot of hard physical labor.

Plumbing lines may be blocked by rust. See that there are valves available to shut off various fixtures in case of leaks and check to see that the valves work. In old houses the repair of a single piece of piping or a fitting may lead to the repair of the entire line because many of the lines (if they are the original ones or have been done over a long time ago) are held together by nothing but rust. Again, the amateur can repair plumbing, but very often parts for old plumbing or heating systems are extremely difficult to obtain. An old house is very likely to have bathtubs, kitchen sinks, and washbasins on porcelain legs and pull-chain flush boxes. These can be seen. Worry about what you can't see. In the old country house be sure it has a modern septic tank and drainfield system. If it has only an old cesspool, you may have to install an entirely new sanitary system. The town code will tell you.

The electrical system of an old house—if it has not been touched in recent years—will require a major overhaul or even a complete replacement. You must not forget that when you start a major rehabilitation of a house you are subject to present-day building codes, of which the electric code is one of the most stringent. There is a very good reason for this inasmuch as electrical work poorly done can cause serious fires. The major electrical work, including the necessary circuit breaker-box and junction boxes, had best be done by a licensed electrician who is familiar with code requirements. You can hang fixtures or even run some wiring to a wall outlet—but if you don't know exactly what you are doing, please don't do it! The consequences can be disastrous.

### 4. THE WOODWORK AND FLOORS

Sometimes the old carved cornices and baseboards or the fluted pilasters and lintels over doorways are the most charming part of an old house. There may also be good wood doors or beautiful mantels over the fireplaces. Some houses even contain old pine paneling which may be covered with as many as twenty coats of paint. Be prepared to use paint remover, scrapers, lye, and a blowtorch (the latter carefully). Be prepared to spend hours on ladders and on your knees first getting the paint off, then putting new paint on. Keep in mind that any molding, cornice, or panel will probably be extremely difficult or very expensive to match.

Many old houses have wide board floors. The newer ones (fifty years or so) may have good oak or other hardwood floors. The wide boards may be badly warped; if so, they will be difficult to replace. The best way to clean them is to rent a heavy-duty electric floor sander, which must be used very carefully. It may be possible to sand warped boards down to a flat surface; if not, the best thing to do is to cover them with plywood and a wall-to-wall carpet—unless you have the time, money, and patience to match them.

Kitchen cabinets in old houses are primitive. If they are in reasonably good shape perhaps you can live with them—after scraping and repainting. In the high-ceilinged old city house you may need a resident ladder to reach the top shelves.

## 5. PLASTER

The interior walls of old houses usually are finished with wood lath and plaster. Over the years the plaster has gotten tired and the wood lath has started to warp or to pull away from the ceiling beams or wall studs. If there has been dampness present some of the plaster may give way at the lightest blow. Removing such lath and plaster and relathing and plastering is a major job. A great deal of the removal is just hard labor and can be done by you, but it is a long, tedious, dusty, dirty job. You will be removing literally tons of old plaster mixed with broken wood lath. You will have to dispose of this rubbish at a town dump or hire someone to haul it away. When you have removed the old lath and plaster you can carefully examine the ceiling joists for any sign of rot and you may be able to reinforce any weak spots.

The replacement of the old plaster can be done in many ways. You can use wallboard with taped joints, a job which you can do yourself fairly quickly. The walls will sound hollow, but to the uninitiated eye they will look finished. If you are a purist, however, you will want to use gypsum lath or metal lath and plaster. If you are going to do this, please practice first in some out-of-the-way corner. Mixing and applying rough plaster is at least semi-skilled work and the application of a white finish coat (even the newer textured ones) is skilled work.

## 6. THE ROOF

You can look for dampness or signs of former dampness on the top floor ceilings or in the attic. Damp spots mean, of course, that there is, or was, a roof leak. If it is a flat-roofed city house, get up on the roof and look for signs of curling and bubbles in the tarpaper or gravel surface. Examine the edges of the roof carefully to see whether the flashing between the roof and the parapet has cracked or curled. Look for curled shingles on a peaked roofed house. Asphalt shingles are good for from fifteen to twenty years; after that they start curling and tearing away in heavy winds. Reshingling a roof, whether wood or asphalt, is not a bad job if you are not afraid of heights. On a flat roof it is best to have the work done professionally with hot pitch and heavy tarpaper. A good roof will last for twenty years.

Examine the gutters and leader pipes. Unless the originals were of copper, they have long since rusted away and have been

# REHABILITATING AN OLD HOUSE

replaced. If they are of anything but copper or aluminum, be prepared to replace them.

## 7. YOUR TIME AND EFFORT

Last but far from least in the rehabilitation of an old house comes your own time and effort. Be prepared to spend almost every waking hour not spent at work or eating or sleeping in working on the house. Be prepared to have everything, including your children, covered with a fine layer of plaster dust, sawdust, or just plain dust. Don't tear anything open to the weather that you can't cover in case of rain, and don't open anything to the weather in the winter.

If you can do so, it is well to make a progress schedule, much as a builder would do for a new house. First put down everything that has to be done. Obviously, you tear out before you replace, but, for example, don't figure on tearing out all the bad plaster in the house at the same time. If you do so you will make every room uninhabitable. Tear out one room at a time and protect the rest of the house with plastic sheets tacked to the doorways. Do the same for other noisy, dusty jobs such as floor-scraping.

If you start doing plumbing work be sure you have all the pipe and fittings you need. Don't start a job you can't finish in a short time unless you are prepared to do without that portion of the plumbing until you are finished.

When you prepare your progress schedule you must think of what comes first in the sequence of rebuilding. For instance, after you tear down old plaster be sure that wiring for electrical outlets is placed before the new wall is plastered. The same goes for plumbing and heating lines.

When you get discouraged, remember the house has been

there a long time and if you do a good job you can enjoy it for a long time to come.

## 8. THE NOT-SO-OLD HOUSE

A house that is about fifty years old will require some work to modernize it, but in general (as was mentioned before) the structure itself should be sound. The heating plant probably will be a steam boiler (probably converted to oil) and steam radiators. You can live with them. Carefully examine the condition of the pipes and valves, some of which may have to be replaced. So far as the plumbing is concerned, look for rust spots around valves and fittings. Turn all valves on and off to see if they work. Try the flow of water in the fixtures. If it doesn't come out full force you may have to replace some piping. Plumbing fixtures may look old-fashioned but you can live with them until you have the time and money to replace them.

The electrical work is different. If you require more outlets and have a number of electrical devices, such as a dishwasher, a clothes dryer, and so forth, the chances are you will have to install more circuits. In doing so you must comply with the new electrical code, and this is a good thing. It may be costly, but how much are your house and your safety worth?

In conclusion: if you buy an old house or a not-so-old house, enjoy it. You can overcome all your difficulties by working hard, not discouraging too easily, and programming your work properly. Good luck!

# CHAPTER 9

## HOUSING DEVELOPMENTS ON ARTIFICIAL LAKES – AND VACATION HOMES

1. An Ocean of Artificial Lakes
2. A House on a Lake?
3. How an Artificial Lake or Pond is Made
4. Embankments and Dams

1. AN OCEAN OF ARTIFICIAL LAKES

Since the early 1930s private landowners in the United States have built over 2 million artificial ponds and lakes.[1] In the beginning small ponds usually were built by farmers as a reasonably reliable source of water during dry spells. They were used for watering livestock, for irrigation, for spraying, for fire protection, and for fish stocking. The U.S. Department of Agriculture was interested in this project and played an important role in the financing and technical expertise necessary to build these ponds.

There are also large lakes which have been created by dams across many of our great rivers. These impounded bodies of water—some of tremendous size—are used for recreation in addition to their primary use for power, water supply and irrigation. Then there are the artificial lakes, large and small, that have come into existence for the sole purpose of adding "romance" to homesite developments.

2. A HOUSE ON A LAKE?

Everyone would like to own a house near a body of water where he can fish, swim, sail or row a small boat, or just picnic or sit by its shore. Shorefront property on salt water, natural lakes, or even large streams has become prohibitively expensive because it is desirable and scarce. This situation has attracted developers and promoters who are using it to their own advantage. Some developers of housing developments or subdivisions

---

[1] A pond is a small body of inland water, often artificially confined. A lake is larger. Many developers call a pond a lake. This chapter will use the words interchangeably.

on artificial lakes try to do a good job but others are simply hustlers, out for a "quick buck." They spread golden tales in the real estate sections of newspapers; they give free dinners for potential purchasers; they use telephone and mail advertising. You may even get an offer of a set of dishes or a dishwasher for just coming to look. All of this is reminiscent of the great Florida land boom of the 1920s when thousands of investors were badly hurt or wiped out.

Therefore, let the buyer beware! The desire for a home on or near a lake is natural and understandable. Some developers offer golf courses, tennis, skeet shooting, and other sports. With today's crowded cities and semi-suburban living, such lures are almost irresistible. It has been estimated that nearly 4 billion dollars a year are being spent on the development of lakeside recreational subdivisions. Unfortunately, many of these have already been spoiled by too intensive use; by sewage from upstream or improper installation of septic tanks; by dam failure; by excessive seepage; and by sedimentation and the growth of water weeds. Many states have passed laws to regulate such subdivisions, and the Federal Government, through HUD, has forced all developers who have fifty or more lots or who advertise across state lines to register their development with HUD's Office of Interstate Land Sales. You have a right to see this report before you agree to purchase. Please do so!

Even if it is a vacation home not on a lake that you want, you must look into it carefully. We all have seen the highly colored brochures and advertisements in the Sunday supplements about "your desert home." *Before you invest, ask:* where is the water and how much; where are the roads; where are the nearest public facilities? It's your money!

## 3. HOW AN ARTIFICIAL LAKE OR POND IS MADE

It is important for the prospective buyer of a house on an artificial body of water to know how it was made. He should also inquire as to how well it was made and under whose supervision. An artificial body of water of any size should be created only with expert engineering, hydrographic and sanitary advice, and supervision. You can obtain information about the lake you are considering from the county or township building inspector or health authority, or from the local agent of the U.S. Department of Agriculture. Don't hesitate to do this. *You may be investing a lifetime of savings in a house on a lake which, through improper construction or inadequate provision for maintenance and sanitation, could be laden with endless trouble and expense.*

Ponds can be created by several different methods. The dammed or embankment pond is made by building a dam or embankment across a flowing stream bed at a point where the slope of the land to the stream creates a body of water at least six feet deep over most of its area. This depressed area is really a shallow valley. If it is too deep it can cause difficulty at the dam site. The site should be carefully selected so that the dam may be built at the narrowest part of the valley through which the stream flows. The slope of the land should be such as to impound the largest possible body of water. This requires surveys to ascertain the exact elevation of the land (contour lines). To make sure that the lake will retain its depth, the watershed area or the stream flow must provide sufficient water.[2] If the development is located east of the Mississippi River or on the Pacific Coast, this presents no great problem because these areas have the greatest rainfall.

The other method of creating a pond is by scooping or digging a depression in an otherwise flat piece of land. Such a pond depends on rain for its water plus some runoff from the surrounding area and on the level of the water table of the

---

[2] A safe figure is five to eight acres of drainage area for every acre foot of impounded water in the middle west; and this number can be reduced to 2 acres further east or on the Pacific Coast. (Rainfall increases as one comes closer to the Eastern seaboard or in certain portions of the Pacific Coast).

ground water. This method can be used only for small ponds. Sometimes a dug pond can be located on sloping land so that it can gain the benefit of the runoff of rainwater. In this case the water must be confined by an earth embankment at the lower side of the slope.

A very important consideration in making a pond and retaining its water is the character of the soil under and around it. Gravel and sandy or coarse soil do not hold water. Rock such as limestone may contain fissures which can drain a pond very rapidly. A mixture of clay and fine soil or even sandy clay soil is best for holding water. You should ask about this. If the original soil in place is not satisfactory, then the builder of the pond should line it with a clay-bearing soil and should carefully and thoroughly compress it with heavy rollers. Heavy equipment or the hauling of clay to a site is expensive and is therefore likely to be evaded by a developer—so that unless you are careful you may find that you do not have a lake for long or in dry seasons.

## 4. EMBANKMENTS AND DAMS

The building of an earth embankment or a dam to hold impounded water is not to be undertaken lightly by an amateur or even by a professional developer-builder without expert advice and thorough investigation of seepage rates, sedimentation, underlying earth strata, rate of stream flow, and many other matters. These subjects are not ordinarily within the average developer-builder's scope. The prospective investor in the project must know under whose supervision such a dam or embankment was built. If a dam bursts and causes damage to property, it is very possible that you are partly responsible. As a member of a community association—most lakeside developments provide that you join an association which becomes responsible for the maintenance of the lake—you are responsible for repairs to the dam or embankment if it leaks and your lake starts to disappear.

## CHECKLIST

*The Questions You Must Ask*

Before you purchase you should ask the developer or yourself the following questions (some of these are in the HUD Office of Interstate Land Sales report mentioned earlier):

1) How was the lake or pond built and under whose supervision?

2) What is the character of the underlying soil? If it is porous, has a clay or other nonporous material been applied to the bottom and sides?

3) If the dam or embankment should burst, is it likely to cause extensive damage? You as a member of the lake association are liable.

4) If the lake is fed by a stream, what is upstream? Are there any sources of pollution, such as a farm with livestock, a small manufacturing plant, or just a lot of houses with septic tanks?

5) What provisions are being made for the disposal of the wastes from the development?

6) Ask some general questions, as you would for any development:

# DEVELOPMENTS ON ARTIFICIAL LAKES

a) Are the roads to and through the project completely developed and fit for all weather?
b) What about schools, shopping, and medical facilities?
c) Is there a reliable source of water supply?
d) How many houses or lots can the developer build or sell and how close together are they?
e) Are there zoning laws to which he must adhere?

If the development is new, look around. The open fields and hills you see now may soon be covered with houses. Is this consistent with your desires for the future? *Are you leaving a crowded area for one which may become just as crowded? Please think carefully before you buy.*

# CHAPTER 10

# ARCHITECTURE AND THE LIVABLE HOUSE

1. An Introduction to Architecture
2. Avoid Momentary or Superficial Styles
   A. The Picture Window
   B. The Split-Level House
   C. The Conversation Pit
   D. The Cathedral Ceiling
   E. The Ever-Present Window Shutter
   F. The Phony Mansard Roof
   G. The Norman Turret and the Tudor Style
   H. The "Olde" English Cottage
3. What Constitutes Good Architecture
   A. Regional Architecture
   B. Effect of Climate and Availability of Materials on Architecture
   C. Discussion and Conclusion
4. The Livable Layout
   A. Room Arrangement for Convenience
   B. Room Arrangement for Your Life Style
      i. For Outdoor Living
      ii. For Your Hobbies
   C. Room Sizes

# 1. AN INTRODUCTION TO ARCHITECTURE

One of man's basic needs is shelter. As he progressed in culture he demanded more than just protection from the elements, thus architecture was born. We can say that architecture is the art of building to secure beauty as well as utility by the arrangement of the mass and plan of a structure. Experts in the field of architecture and construction are disturbed by the fact that good architecture is seldom found in the single-family house. This is unfortunate because good architecture costs no more than poor or mediocre architecture. However, an informed public can do something about this situation. We have all heard the expression, "I know what I like," whether it be about books, plays, art, or architecture — but do you? It is possible that, given more choice, you might like something else better. This chapter will try to inform you about good architecture.

Before we go any further, please be assured that this is not an appeal to architectural snobbery. It is simply an explanation of how the proportion, mass, and interior space of a house can be arranged so that it never will be dated or out of style.

The author lived in a housing development during World War II. It was near several large manufacturing plants and the housing shortage was so acute that anything built was immediately rented or sold. The builder-developer, because he did not have to attract tenants or purchasers, built small houses with simple lines and no attention-getting or superficial decorations for sales appeal. He planted a tree in front of each house and curved the roads slightly so that the houses did not look as though they were in a rigid row. At the present time (almost thirty years later) these houses could have been freshly built and command top prices in their category. A house with good lines and good plan will always be pleasing to you, as well as being a good investment.

# ARCHITECTURE AND THE LIVABLE HOUSE 103

## 2. AVOID MOMENTARY OR SUPERFICIAL STYLES

The use of "gimmicks" in present-day small house design is an attention-getting tactic of no lasting value. Instead of building houses of good materials and proportions, many speculative builders resort to such devices in an attempt to sell their houses. These fleeting styles have no architectural reason for being and therefore do not last very long.

### A. THE PICTURE WINDOW

For a time the picture window was a feature of every speculatively build house, and even today this gimmick still is being widely used. If the area around the house gives it sufficient privacy, then large glass areas are an attractive feature in a contemporary house. They blend the surrounding exterior with the interior to create a very pleasant mode of living. A large window overlooking anything of scenic interest is a very desirable feature in a house; but a picture window overlooking a neighbor's picture window across the street just doesn't make sense. The large glass area can impair your privacy as well as requiring extra radiation in the winter and making the room hard to keep cool in the summer. It also presents a furniture placing problem, as well as requiring special Venetian blinds or drapes.

### B. THE SPLIT-LEVEL HOUSE

Another style or device is the split-level house. There is a valid reason for such a house on a sloping lot. By splitting a house into three levels with a half-story height difference between them, the house rises with the slope of the land and every part of it has full exposure. When a split level house is placed on a level lot, however, part of the house must be partly buried in the ground and become sort of a semi-basement (see Figure 8), or the builder has to create an artificial hill for part of the house if he wishes to keep all three levels above grade. The split level also costs more per square foot of livable space. You would build a split-level house on a level lot if you wished to provide more house on a confined lot without having to build a two-story house. Some people dislike climbing stairs so much that they will settle for this.

## C. THE CONVERSATION PIT

One new style in recent vogue is the "conversation pit." This is simply a space in the living room a few steps below the general level of the room—apparently on the assumption that this design somehow promotes conversation. It is expensive to frame; the difference in level is a hazard; and except for use in a chichi Hollywood set it makes no sense architecturally.

## D. THE CATHEDRAL CEILING

Many of the higher priced speculative houses now advertise a "cathedral ceiling." The two-story, or even higher, ceiling in a living room can produce a striking architectural effect in a house designed for a country estate or a hunting lodge, where the room is large enough to be in proportion to the extra height. To build an ordinary sized living room with a high sloping ceiling is a waste of space and can detract from the charm of a house. One finds the eye wandering up into the darkness of the upper spaces. If the room is not very large it will look as though it is at the bottom of a well. The high ceiling requires extra insulation in the summer and heating in the winter. It does not add to the salability of a house and may, in fact, reduce the number of possible purchasers. Again, however, it is

FIGURE 8

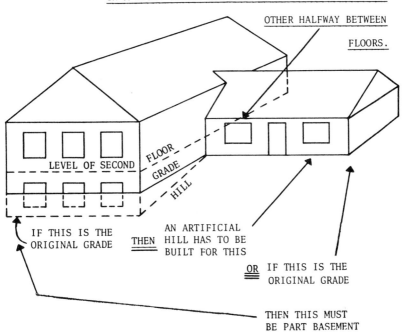

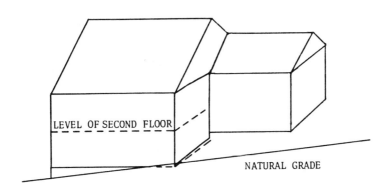

your choice. If you are charmed by a high-ceilinged living room, by all means buy it.

### E. THE EVER-PRESENT WINDOW SHUTTER

One of the longest lasting styles in exterior house decorations in this country is the window shutter. The original purpose of the shutter as a protection against hostile Indians or severe weather (before the advent of central heating) is no longer valid. Shutters still are used extensively in many countries as a protection against the heat of the day and even as a guard against intruders or the imagined ill effects of night air. They also serve as protection against tropical storms and can be used as safeguards for seasonally used houses. But tacking shutters to the side of large windows or on either side of a door, when the most cursory observation will show that the shutters are not made to close and would not fit anyway, is not good architecture. Shutters are difficult to paint and if not used may harbor birds' or wasps' nests. You can spend your money more usefully on other things.

Perhaps this is too severe an indictment against the shutter. Many older houses have shutters which served a useful function and many modern Colonial or New England style houses use shutters as a decorative feature. Let us say, then, *don't put shutters on a house unless you feel they are absolutely necessary for decoration and you think your house looks bare without them.*

### F. THE "PHONY" MANSARD ROOF

The mansard roof was a device for getting an extra floor on a house and yet making it look as though this extra story was a gracefully sloping roof (see Figure 9). Because of the steep slope of the mansard, one can obtain a floor area nearly as large as the floors below, and there are many older houses which have used this design feature to great effect as a roof line above a two- or three-story height. Lately, however, builders of townhouses and even single-family detached houses have been placing a mansard roof over the first story and using it as the exterior wall of a second story. They do this because they get a roof and a second story at the same time. The effect is that of a hat pulled over one's eyes and ears, and the house tends to look squat and grotesque. Builders are always experimenting and some people (to their later regret) pay for these experiments.

G. THE NORMAN TURRET AND THE TUDOR STYLE

These styles are not prevalent now but they have been in past years. They date a house back to the early 1920s. They are most often seen in good, established suburbs. The turret is usually in a corner and the front entrance is located in it. It provides a lot of unusable space at a high cost.

There are many excellent examples of the Tudor style in our better suburban areas. With its fine bow windows, ornate chimneys, half timbering, and patterned brickwork, a good Tudor house can be a thing of beauty. Unfortunately, such houses are quite expensive and the style dates the house back at least to the early 1920s. What is more, this style has been imitated by cheaper builders with disastrous results.

H. THE "OLDE ENGLISH" COTTAGE

This is another eye-catching style which has no permanent architectural value. It uses diamond-paned windows in some places and half-timbering which, if not done well, is difficult to maintain. Diamond-paned windows and other details of the cottage style are now coming back in fashion in some parts of the country, notably the Pacific Coast where it is known as the

FIGURE 9

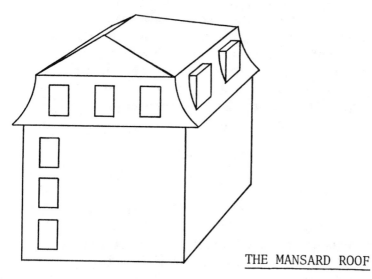

THE MANSARD ROOF

THE ORIGINAL MANSARD ROOF WAS DESIGNED TO ADD AN EXTRA FULL STORY TO A HOUSE IN ADDITION TO GIVING IT SOME ARCHITECTURAL INTEREST.

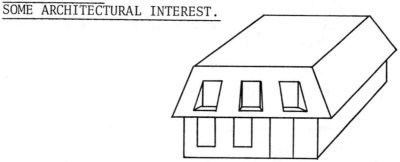

THE IMITATION MANSARD. THE FLAT ROOF, THE RECESSED WINDOWS AND THE SLOPING SIDES OF THE SECOND FLOOR MAKE THIS AN UGLY, UNCOMFORTABLE HOUSE.

# ARCHITECTURE AND THE LIVABLE HOUSE 109

"Cinderella Style." Some of the development houses in this style even have built-in birdhouses or dovecotes over the garage. If you like it, buy it. The style may not last, but it looks "cute."

Development builders, in an effort to make a quick sale, are constantly trying eye-catching new tricks. Let your own good taste be the judge.

## 3. WHAT CONSTITUTES GOOD ARCHITECTURE

To the individual buying a house, architecture is a very personal thing. Your home is more than a protection from the elements. It affects every aspect of your life, influences your children's present and future, and reflects your taste. No one should tell another what kind of house he should like and purchase, any more than one person can tell another what kind of clothes to wear. Personal likes and dislikes are very important.

There are basic rules, however, which may assist you in making a lasting satisfactory choice. Consider architecture as a social art because it shapes your physical environment and because the line and shape of this environment is very important to your well-being.

Your house not only should mirror your likes and way of living but also should reflect its environment. The materials of which it is made should, to some extent, conform to the kind of materials which are available and generally used in the area. Its size and shape should be pleasing to the eye in relation to the size of the building lot, the character of the land, the view, and the privacy that is available. In other words, it should be in sympathy with the land on which it is built and with the character of the land around it. Its architecture can be contemporary or traditional but it should not clash wildly with the architecture of the general area. Think of a slab-sided modern house or a Japanese teahouse on a street that is lined with good Colonial or

Georgian houses. Let your personal taste be the judge on this matter.

## A. REGIONAL ARCHITECTURE

As an example of regional design one thinks of the traditional Colonial architecture of New England—the Cape Cod house, or salt box—with its steeply pitched roofs and comparatively small glass areas. The pitched roofs were built to shed heavy snow and the windows were small because glass was very expensive and hard to get. It is also a poor insulator against extreme cold. The windows had small panes because glass in large sizes was not available. The houses were two-storied because twice as much roof would be needed to cover the same area if it were built with one story, and roof areas let heat escape and leaked in heavy winter storms. A two-story house is also easier to keep warm because heat rises and the house could be planned around a central fireplace or, in later times, a warm air register. Because of severe winters, the houses were connected by covered paths to barns or workshops.

Although these houses were built for that period, they are as valid today as they were then. Their simple lines and handsome proportions make them ageless. If you like such a house you can do no better than to buy or build one. This does not mean, however, that you should not buy or build a flat-roofed contemporary house with a large glass area in New England. If you want such a house, and if it is located on land with a view and with privacy from road and neighbors, and if it is in harmony with the land around it, by all means buy it or build it. But not on a small flat lot surrounded by other houses.

Another regional design is the low rambling ranch house of the Southwest with its long low roof line. Originally the walls were thick to keep out the heat and the roofs were of baked clay tiles. This style has now been adapted throughout the country but is still most common in California and the South-

## ARCHITECTURE AND THE LIVABLE HOUSE

west. The roof can be of heavy wood shingles or shakes, although the orange-red Spanish tile is still an authentic and beautiful feature. The overhanging eaves form a shield against the sun. The walls can be of redwood or native pine. The garage can be a simple carport. The entire design is for a house in a mild climate with much outdoor living and with no severe changes in the weather. The one-story house is also suited to an area which is prone to earthquakes. Properly designed, the ranch house can be built anywhere. It is a very convenient house and its long low lines can be most attractive.

Another Southwestern style is the "Territorial," which uses concrete block and stucco or adobe blocks. The thick masonry walls provide good insulation. Many such houses are being built without furring (*see* Chapter 5-5b) on the inside. If you can afford the modest extra cost involved, have your house furred. It is a good investment.

There were many other styles of architecture which were built to serve a purpose. For example, we had the Louisiana plantation house with its railed verandas and large windows and the Greek Revival Southern plantation house with its two-storied columned veranda. These stately houses were built to impress residents of the area with the wealth and importance of the owners. They had to be large because they were the center of social activity for many miles around. They were very often built on a knoll (real or artificial) to enable the planter to overlook much of his land. He could truly say, "I am monarch of all I survey."

To build a house at the present time on a small plot, with a colonnaded front to resemble a Southern plantation house, is the height of affectation.[2] Yet development builders do it and occasionally people buy these houses. If they remained unsold, the builders would soon stop building them.

---

[2] A small plot in this case can be any plot less than two hundred feet wide. Depending on the surroundings, such a house may not be suitable on even wider frontage.

## B. EFFECT OF CLIMATE AND AVAILABILITY OF MATERIALS ON ARCHITECTURE

The regional styles mentioned in the preceding section of this chapter—e.g., the plantation house, the ranch house, the thick-walled Territorial or Southwestern house, the New England or Cape Cod house—have been developed because of climate and geography as well as the customs of the region.

In Northern regions where there is a good deal of cold weather and snow, the roofs are steep and the window openings are small. In present times, with excellent central heating available, window areas can be as large as you want, but if you do have large glass areas, consider your heating and/or air conditioning cost. Such Northern houses usually are built of readily available wood and have insulated walls and hardwood floors.

The house in the milder semi-arid climate of southern California need not be insulated. In this area the average house is built of wood studs covered by heavy felt paper to which is nailed chicken wire. This in turn is covered with stucco which can later be tinted. Because there is no snow load, the roof needs to be only slightly pitched or even flat and can be of lighter construction. Redwood is much used because of its availability. Floors can be of ceramic tile or resilient tile. You must be sure that the stucco has a cement base and that the supporting wire is carefully fastened to the structural frame. Southern California houses of better quality or Northern Pacific Coast houses will be built of wood sheathing under the stucco. Because of the availability of heavy wood shingles or shakes, they are frequently used for walls and roofs.

In the Southwest, where wood is not a local product, houses frequently are built of concrete block or adobe block, which, depending on the quality, can be cement coated on the outside and furred on the inside. Because of the heat the walls are thick. There is very little rain, so the roof can be flat and built of tarpaper, hot pitch, and gravel. (All flat-roofed houses, anywhere, should have such roofs.)

In areas where hurricanes or tornadoes are prevalent, you are not allowed to build a wood frame house.

# ARCHITECTURE AND THE LIVABLE HOUSE

In regions where clay or shale brick is readily available, the private home may be built of brick, either as a solid wall or as a facing over a structural wood wall.

In all climates where the weather is hot in the summer and cold in the winter, walls and roof should be insulated. This applies especially to flat roofs.

The prospective builder or purchaser of a home is urged to speak to local builders or architects and to investigate the availability of materials and the architectural style that best suits the environment. As previously mentioned, anything you want is permissible anywhere you want it, so long as it is consistent with the local building and zoning codes. But before you decide, please investigate the customs of the area.

## C. DISCUSSION AND CONCLUSION

It is a fact that the public clings stubbornly to what it is used to. Custom and tradition have thwarted architects, who always find that they have to fight for modern design. It does not matter whether the house you purchase is contemporary or traditional; whether it has a flat roof, a shed roof, or a peaked roof; whether its exterior walls are built of cement block or wood or native stone or are all glass. What does matter is that the house fits its environment and your way of life. It must be pleasing to you but at the same time its size and shape and building materials should be in harmony with its surroundings. *The supreme test of good taste and good architecture is for your house to look as though it belonged.*

## 4. THE LIVABLE LAYOUT

Many informed people say that the size and shape of a house must be products of the interior design and plan. In other words, the developer-builder, the speculative builder, or you—if you plan to build your own house—should decide on the number and shape and the size and layout of the rooms, and then wrap the outside around them. This is only partially true. The exterior and the interior should work together. There must be compromises. Whether you build or buy, you should look for a house in which the functional arrangement of the interior space meets your needs and your mode of life and which, at the same time, meets the requirements of good architecture. This is not as difficult as it sounds. Please do not expect to find your "dream house" all ready and waiting for you. If the house meets your basic needs and the room arrangement is reasonably convenient, a little ingenuity on your part will go a long way toward making it your home.

## A. ROOM ARRANGEMENT FOR CONVENIENCE

A number of simple rules regarding the size and shape of various rooms and their relationship to each other have been proven, by experience, to make daily living more convenient. Today every house must provide shelter for at least one automobile. In mild climates such a shelter can be an open carport which is roofed over to protect the automobile from sun, dew, rain, or snow and to protect you when you are getting in or out of the car in bad weather and your arms are full of groceries. If at all possible, therefore, the garage should be attached to the house by some overhead coverage. Although a garage or carport is necessary, it does not need to be highly visible. Judicious placing of a driveway can put it out of sight of the front entrance.

When you arrive home from work or shopping, or when the children come home from school or play, you should all be able to enter your house through a work area. This can be a kitchen, a laundry, or a "mud room." This is where you can leave your wet rubbers or umbrella and the children can shed their muddy boots. This is where you can dump your groceries or bundles before entering carpeted areas. The stairway to your basement area should also be in this work area. This is particularly convenient if your furnace room or hobby shop are located there. Repairs can be made and material brought into the house without danger of spoiling rugs or banging furniture. (see Figure 10.)

Your living room should be immediately adjacent to the front or main entrance. This is where your guests will arrive and be entertained. If at all possible, the living room should not have to serve as a passage to other rooms. If the house is large enough for a center hall, then this hall can serve as a means of passage to all rooms. If there is no center hall, then it may be possible, in a two-story house, either to place the stairway to the upper rooms at the end of the living room closest to the front door, or, in a single-story house, to have the entrance to the other rooms along one end of the living room. (see Figure 11.)

Whether your house is one-story, two-stories, or split level, your bedrooms should open into a common hall or passage which also gives access to a bathroom, unless, of course, each bedroom has its own bath. In this case, your guests can use one of these bedroom baths unless you provide a powder room (basin and water closet). Try not to have more than one entrance to any bathroom, for this can lead to embarrassing situations. If a bathroom is for the use of the occupants of more than one bedroom, its door should open into the hall.

If you have a dining room or a dining space as part of the living room, your kitchen should be adjacent to it. Being able to pass from the kitchen to a bath or bedroom without going through the living room is a convenient arrangement. If guests are present and the lady of the house is doing the cooking, it is very nice for her to be able to slip off to powder her nose before announcing that "dinner is served".

Many homes now have a family room for children and games, watching television, and relaxing. It should be near the kitchen and work areas and away from the bedrooms.

If the house is not exactly what you require, you must be prepared to change the use of rooms to suit yourself. Of course, a kitchen or bath or laundry must stay where it is, but nobody

FIGURE 10

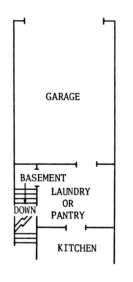

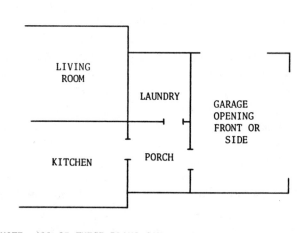

NOTE: ALL OF THESE PLANS CAN BE REVERSED AND GARAGES CAN BE FACED WITH OPENINGS ON ANY SIDE EXCEPT WHERE ATTACHED TO HOUSE.

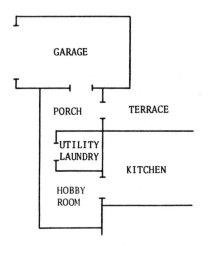

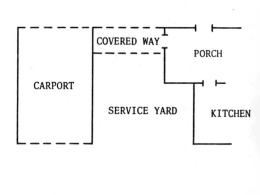

WAYS OF ENTERING A HOUSE FROM GARAGE TO WORK AREA

FIGURE 11

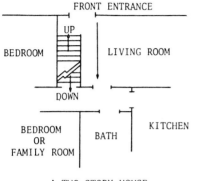

WAYS OF ENTERING A HOUSE FROM FRONT ENTRANCE WITHOUT GOING THROUGH LENGTH OF LIVING ROOM (IN ABSENCE OF CENTER HALL)

A TWO-STORY HOUSE

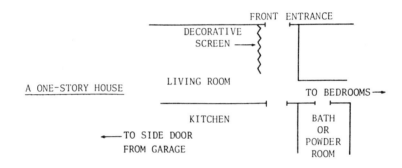

A ONE-STORY HOUSE

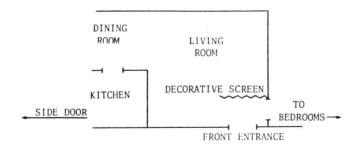

A ONE-STORY HOUSE

says that you can't make what was formerly a living room serve as a family room and vice versa, or use a bedroom as a study or den or library or whatever you wish.

### B. ROOM ARRANGEMENT FOR YOUR LIFE STYLE

Your way of life depends to some extent on the climate, the sports you enjoy, your hobbies, your children, your work, and your likes and dislikes.

1. For Outdoor Living

Some people like to be outdoors as much as they possibly can. If the weather allows, they eat and spend all their leisure time outdoors. If you are such a person, you should try to find a house with maximum access to your lawn or terrace or porch. This can be done by direct access from your kitchen or through a breezeway-porch adjacent to the kitchen, or by direct access to the outdoors from your dining room or even a bedroom. Such outdoor living depends to a great extent on the climate. In a cold or rainy climate you may have to do your outdoor living in a sheltered spot. If you are building for yourself it is not as much a problem as if you were buying a completed house, which may already have a terrace or patio or porch. Perhaps, then, you can install a windscreen, awning, glass doors, or wide-opening windows on a covered porch. A porch can be a fine addition to your everyday living.

If you are an enthusiastic gardener, try to find a house that has a sunny patch of ground near your indoor work space so that you can pick up and leave your tools, fertilizers, and so forth in this area. You may have room for a tool or storage outhouse. Some people have been known to build a small glassed-in area or miniature greenhouse in a sunny spot close to the house.

The same considerations should be applied to any outdoor sports that you may like and to the location of a children's play area. Such areas should be near the kitchen, laundry, or porch so that muddy children or hot adults can wash or remove their shoes before tracking up the house. Having the play area near your work area is also important for supervisory purposes. It is very helpful if you can see what is going on while doing your work.

2. For Your Hobbies

If you are bridge or card enthusiasts, or if the man of the house likes to make things, or if you like to read and someone else wishes to watch television, it can be done by using rooms for multiple purposes. A previous section of this chapter discussed the arrangement of rooms for convenience. Almost all houses either have an attached garage or one within a few steps of a back door. The man who likes to make things can set aside some room for himself in this garage, in the basement, or in the family room. If there is no machinery involved, an extra bedroom can be used as a shop, a sewing room, or a study. Many houses do not have a family room; this type of room is relatively new and, in any case, it adds considerably to the cost of a house. If there is no family room, the living room or a basement can serve as one, and skillful use of a kitchen, dining alcove, or dining room can separate such noncompatible activities as reading and television-watching. If the rooms are arranged for convenience you will find that they can be used for your hobbies as well, without too much dislocation of your ordinary living.

## C. ROOM SIZES

A house need not have very large rooms to make it livable and comfortable. There are many houses with large living rooms which look almost barnlike because they have not been furnished or decorated properly; and there are houses with comparatively small living rooms which look cozy and lived in. There are, however, certain minimum sizes which should serve as guidelines. A living room should be at least thirteen feet wide in order to enable you to have seating on both sides of it and still maintain a useful clear space in the middle. It should not be much less than twenty feet long for the same reason. If it is smaller than this it begins to look cluttered.

Bedrooms should be not less than ten feet in either direction. By the time you place a bed, a dresser, and a bed table in a bedroom and have doors opening into closets, the ten feet will be barely adequate.

Bathrooms should be at least seven feet wide so that you can walk between the tub and the basin and the water closet without bumping into anything.

Your kitchen can be short and wide or long and narrow. Its convenience depends on the location of the closets and the appliances. (There will be a discussion of kitchens in Chapter 11.) If you would like to eat most of your meals in the kitchen, try to find one with a space large enough for a table and some chairs. The stools and cute little counters for kitchen eating that are shown in some decorating magazines are in most cases quite uncomfortable.

If your house has a large pantry, a dressing room between bedroom and bath, a larger than usual breezeway, porch, or garage, or even a family room or a dry basement, you can utilize these areas for more comfortable living.

In this day and age a two-car garage becomes almost a necessity. If you can possibly do it—build a two-car garage!

The conclusion that may be reached from all the foregoing suggestions, discussions, and cautions is that good architecture can provide you with a house which will be a pleasure to live in and a joy to look at. A well proportioned house with simple lines which fits its environment can give you a constant pride of ownership.

## CHECKLIST

*The Livable Layout*

First, two important questions:

*Does the house seem to fit its surroundings?*

*Does it give you a feeling "This is where I want to live"?*

1) Are there enough rooms to meet your needs? Please note that you can use rooms for many different purposes. You must, therefore, base your room count accordingly.

2) Can you get from your car-storage area to the house while staying under cover?

3) Can you come directly into a work area from your car or the outdoors?

4) Can you get to bedrooms or other rooms from the work area or the front door without going the length of the living room?

5) Does the house have a room which is suitable for television, games, or other family activities?

6) If not, can you make the living room or another room serve part or all of this purpose?

7) If you are an outdoor hobbyist, do you have reasonably free access to the outdoors?

8) Check the room sizes carefully and try to arrange furniture in them (on paper, of course).

9) Check the efficiency of your work areas. (*see* Chapter 11.)

# CHAPTER 11

## THE HEART OF YOUR HOUSE: WORK AND UTILITY AREAS

1. The Importance of These Areas in Everyday Living
2. The Kitchen
    A. Remodeling an Existing Kitchen
    B. The Kitchen in a Speculative House
    C. Planning a Kitchen in a New House
3. The Bathroom
    A. Remodeling
    B. Suggestions and Cautions about Bathrooms
    C. The Bathroom in a New House
4. The Laundry

1. THE IMPORTANCE OF THESE AREAS IN EVERY-DAY LIVING

The kitchen is the heart of the house. Food and shelter are basic to life and the preparation of food has always been a fascinating household activity. From the pit of the earliest cavemen to the twentieth-century thermostatically controlled oven, the hearth has symbolized home. For a long time, the large family kitchen was where most of the intimate family activities occurred. Therefore, when buying, building, or altering your house give careful attention to the size and layout of your kitchen, for convenience in the preparation of meals as well as a place for family gathering.

There are other utility areas in your home such as the bathrooms, the laundry, and the furnace room. All of these areas can serve several purposes if proper thought is given to their size, location, and the arrangement of facilities such as light, heat, plumbing, and electricity.

2. THE KITCHEN

Before you plan a new kitchen or alter or remodel an existing one, you must ask yourself a number of questions about your special interests and requirements. Are you a gourmet cook? A baking enthusiast? Do you have a large family, some of whom have meals at odd times? Do you like displays of kitchen utensils? Do you want your kitchen to be a center of family activity? You must also consider your likes and dislikes as to color and architectural style.

A. REMODELING AN EXISTING KITCHEN

If you are remodeling an existing kitchen you may be able to gain space and efficiency simply by rearranging your work space or by closing or relocating a doorway. A large kitchen is a very desirable thing to have, but this is not always possible in an older house without resorting to major alteration. However, if you have a laundry room, "mud room," or pantry next to your kitchen, you should consider breaking through and combining it with your kitchen. Although the ingenious housewife can make do with a kitchen as small as eight feet by twelve feet, a kitchen twelve feet square is better and one twelve feet by fifteen feet is luxurious. There are several ways in which the basic working area can be designed. Its efficiency depends on the size and shape of the room and the location of the windows and doorways.

Let us consider some examples: In an $8' \times 12'$ kitchen you can place a large sink, a four-burner range plus an oven, a refrigerator-freezer (fifteen cubic feet) and a dishwasher, and still have seven feet of counter space and more than ten feet of cupboard space. (see Figure 12.) In addition, there will be room for a small table or counter for breakfast and other informal meals. This kitchen is arranged in an L shape. The same kitchen appliances can be arranged in a U shape or as a work space on either side of the room (like a ship's galley)—or only on one side, depending upon the location of the doors and windows and whether it is possible to relocate them. Many people prefer a U-shaped kitchen with the stove and refrigerator across from each other with the sink in the middle. It seems to be an unfailing rule of kitchen design for the sink to be under a window.

In the kitchen shown, if one of the eight-foot walls separates it from a laundry or utility room, it is suggested that you consider the removal of all or part of the wall. This will give you an opportunity to provide a peninsula counter, a wall oven, or more dining space. The larger you can make your kitchen, the more attractive it will become as a gathering place for your family and friends.

B.  THE KITCHEN IN A SPECULATIVE HOUSE

So far the discussion has been about what you can do with an existing kitchen. You can do a great deal, but when you build a

FIGURE 12

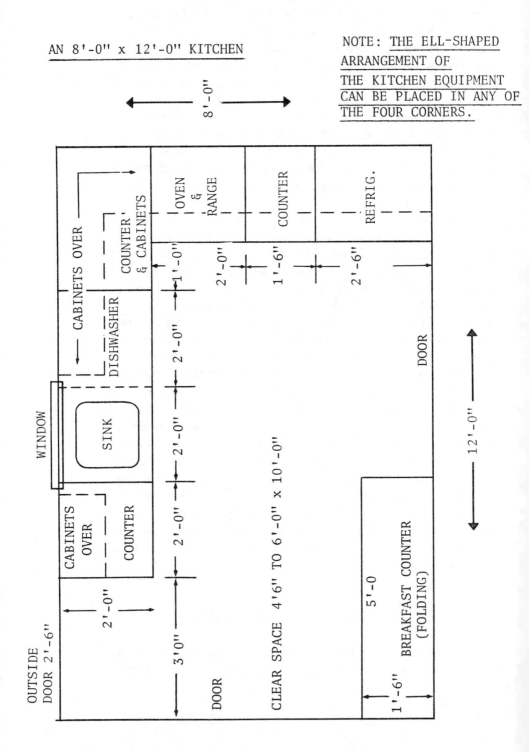

# THE HEART OF YOUR HOUSE

house (and sometimes when you buy a house under construction from a speculative builder) you can do anything—in fact, you can have exactly the kitchen you want. This depends, of course, on how nearly completed the speculative house is. You may be able to change door locations or even window locations if it is just being framed, but it will be difficult to change the room size. Before you change anything, find out what it will cost, not only for the rough carpentry and cabinetwork but also for the plumbing and electrical work if there is any in place which must be rearranged. If it is too expensive to make all the construction changes, you may be able to make some of them, such as relocating a doorway or leaving out a wall between the kitchen and a pantry or entry; or perhaps you can choose better appliances or have them located in different places than the builder had planned. Many developers make a selling point of the kitchen equipment in their houses, and in more expensive ones they furnish built-in ovens, kitchen exhaust fans, large refrigerators and freezers, dishwashers, natural wood cabinets, and sometimes expert layout service. Such new houses usually have family rooms adjoining or near the kitchen.

## C. PLANNING A KITCHEN FOR A NEW HOUSE

We now come to the kitchen in the house you are planning to build. Although the opening sentence of this chapter refers to the kitchen as one of the most important rooms in the house, it should not be so elaborate or so large that it takes a disproportionate part of your building dollar. Indeed, you may not feel that the kitchen is as important as the author does. You must, therefore, decide what your priorities are. If you want a large kitchen-family room, you must decide the maximum size it can be without taking too much from any other rooms you need for daily living. Try several standard dimensions such as 12' × 12' or 12' × 15' or 15' × 15'. See if you can fit all the equipment you want into these spaces. Draw a plan of the kitchen to ¼-inch scale and cut out cardboard shapes to scale for your appliances. In a U-shaped kitchen, place the sink at the window and the refrigerator and stove on either side of the room if you can, or at least separate them with counter space so you can take things off the stove or out of the refrigerator and set them down right away. Plan enough counter space for toasters, blenders, mixers, can opener, telephone, and so forth, and sufficient electrical outlets to accommodate them. (A continuous raceway with an outlet every twelve inches is recommended.) When you are satisfied with your layout and the efficiency of your working space, see how much room you have left for informal eating.

If you can afford it and want to indulge in a conversation piece, you can have a handsome kitchen. You can install a fireplace with a raised hearth and Dutch oven; you can display a beautiful array of copper utensils; you can have it color-planned. It is possible to have an authentic Colonial, Provençal or Spanish decor. It can be a combination kitchen, laundry, and family room separated by a mix-bake center or a planning center. And even with all of these features, it need not be a huge room if it is skillfully planned.

### 3. THE BATHROOM

The kitchen and its hearth have been with us for a long time, but the indoor bathroom is a comparatively recent development. Central heating and sanitation have progressed along with the art of building, and have enabled bathrooms to become more and more elaborate. The bathroom now frequently doubles as dressing room and boudoir, in addition to serving its original sanitary functions.

#### A. REMODELING

Remodeling an existing bathroom requires even more skill than planning a new one. Bathrooms in many older houses are likely to be small without space for accessories such as double basins, dressing tables, shower stalls, or extra partitions. Enlarging a bathroom is expensive because the bathroom usually is between two other rooms and in most cases its longer dimension runs parallel to these rooms, so it can't be widened without taking space away from them. What is more, alterations can prove difficult and costly if the plumbing has to be changed. Lengthening a bathroom is ridiculous if it has to project from the exterior wall or into hall space. This should be done only in conjunction with a major remodeling job.

The minimum space for a practical bathroom is five feet by seven feet. Not much can be done with this room except to retile it, change the basin to a cabinet enclosed unit, and replace the toilet bowl with a modern one. If you have a seven-foot width by an eight-foot depth, you can make changes such as putting a partition between the toilet and the rest of the bathroom or installing a stall shower or a dressing table in combination with the basin. And if the bathroom is nine feet long, then you can have the luxury of a corner tub. (*see* Figure 13.)

Even if you find as little space as 2½ feet wide by 5½ feet long, you can install a toilet and basin on your first floor. With these minimum dimensions you will need either a sliding door or a door that opens out of the room. (*see* Figure 14.) Try to have the door open in such a way that a person using the toilet is shielded momentarily by the opening door.

B. SUGGESTIONS AND CAUTIONS ABOUT BATHROOMS

1) Try not to place a bathtub at a window if you can possibly avoid it. You can swing the tub around so that one of its short sides comes against the exterior wall with the window to one side. If you cannot move the tub, then change the window opening so that the sill is at least four feet and preferably more above the floor. The bathtub is one of the most dangerous places in the house and a slip can result in pushing your arm through a pane of glass or even a fall through a window.

FIGURE 13

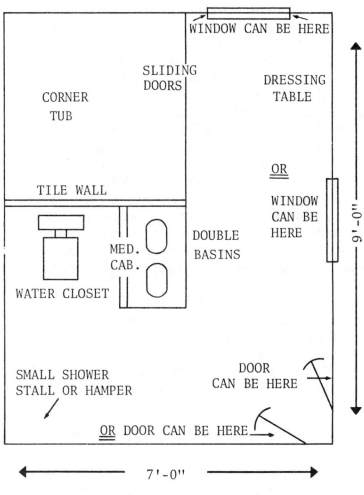

NOTE: THERE ARE ALTERNATIVE LOCATIONS FOR DOOR AND WINDOW

A 7'-0" x 9'-0" BATHROOM

# THE HEART OF YOUR HOUSE

2) If the bathroom is visible from other rooms, at least have the door open (inward, of course) so that if it *is* ajar one does not see the toilet. There are more attractive things to look at.

3) Please place the toilet far enough from the door so that anyone using it is not struck by the opening door. If you can't provide the door with a clear swing, then place the basin or bathtub there.

4) Have towel bars and soap dishes with a grab or grab rail, anchored firmly into the wall. Consider a non-skid surface for the bathtub.

5) Be sure that your stall shower doors or any clear or translucent glass partitions or tub enclosures are of tempered or shatter-proof glass or plastic. *Don't use glass accessories.*

6) Your basin and toilet should be of ceramic china and your tub of enamel fused over cast iron or steel. If the tub is of lightweight enameled steel, be sure that sound-deadening material is packed under and around it before it is installed. If not, it will sound like a steel drum when you are taking a shower.

7) Ceramic tile walls and floors are the longest lasting but there are many other satisfactory materials. You can purchase asbestos- or aluminum-backed sheets of material that look like tile and are waterproof. Make sure your waterproof wall covering goes all the way to the ceiling around a tub with a shower. You can also use plastic-coated paper on walls and ceilings. Floors can be of vinyl or linoleum. Consider carpeting that can be picked up and washed.

## C. THE BATHROOM IN A NEW HOUSE

The bathroom in a house you are building can be made quite appealing and even luxurious without major expenditure. First, give yourself enough room. Consider a seven foot by eight foot dimension in at least one bathroom and don't have anything smaller than 6½ feet by 7 feet in any complete bathroom. Try to make your powder room or downstairs toilet large enough (3 feet by 6½ feet) so that you can swing a door inward.

FIGURE 14

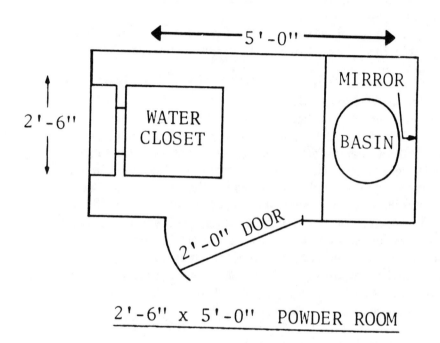

2'-6" x 5'-0"  POWDER ROOM

FIGURE 15

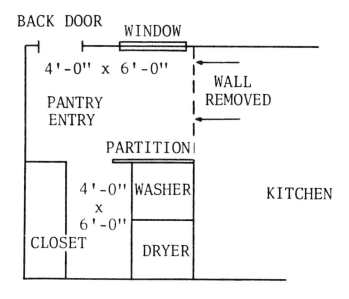

COMBINATION LAUNDRY, MUD ROOM, PANTRY,
WHICH CAN BE USED AS PART OF
THE KITCHEN-BREAKFAST AREA

In your master bathroom you can have double basins, a shower stall, a tub, the toilet concealed behind a partition, and, if the room is seven feet by nine feet, a luxurious enclosed corner tub and a dressing table. Lay everything out to scale before you decide on what you want and where you want it. If you have room, install a narrow closet with doors and seven- or eight-inch wide shelves to use for bath supplies, medicine, sprays, and so forth. You will find this very handy. Also be sure to provide enough electrical outlets for shaving, massagers, toothbrushes, and other electrical devices. One other caution: all interior bathrooms *must* be connected to the outside by a metal duct and an exhaust fan.

## 4. THE LAUNDRY

In older houses the laundry consisted of one or two washtubs situated in the basement next to the furnace room. This was not a comfortable place; the light was either from a tiny window or artificial, and the area was apt to be too hot in winter and too damp in summer. Besides this, dirty and wet clothes had to be carried down and up stairs, making Monday washday not a day to look forward to. The modern housewife will have none of this if she can help it.

Modern laundry equipment such as washer and dryer are now used at any time—indeed, several times daily in large families. The laundry area, therefore, should have adequate light and air and be conveniently located, preferably somewhere near your kitchen. In an older house the pantry can be converted into a laundry and it can serve as a combination laundry-mud room. The sorting and washing of soiled laundry should be done as far as possible from the preparation of food; if you break through a wall, you should keep this in mind. For instance, in Figure 15, which shows one end of an eight foot by twelve foot kitchen, a pantry or entry—if there were one—would almost surely be on the eight-foot side; and this is the usual arrangement in older

# THE HEART OF YOUR HOUSE

houses. It is, therefore, not too difficult to place a low partition or a high counter across this room with one part used for the laundry and the other for the kitchen.

If you can't move your laundry out of the basement, you can at least have a good light and other amenities. You can paint the walls a light color, you can install a mirror over the tub, you can install a counter top and shelves for washday materials, and certainly you can arrange some ventilation, even if it is only a fan. If you can partition it off from the furnace so much the better. Take a radio down with you! Leave a playpen there for the baby and toys for the older children.

## CHECKLIST

*The Heart of Your House: Work and Utility Areas*

*The Kitchen*

- Emphasis on your favorite enthusiasm or activity
- The layout (cut out cardboard models at $\frac{1}{4}'' = 12''$ scale)
- Size of refrigerator
- Freezer (if you want yours in the kitchen)
- Kitchen sink (single or double)
- Dishwasher
- Stove (upright or conventional)
- How many feet of counter space
- How many feet of dish closet space
- How many feet of pot and pan space
- Grocery storage space
- Special storage areas—for infrequently used appliances, trays, large bowls and platters, cooking tools and utensils of all kinds
- Silverware and knives
- Cannisters and spice racks

Do you want an island counter—built-in chopping block or blocks?
Would you like an open fireplace?
Are you the decorative-display-of-utensils type?
Would you like to eat in the kitchen? All meals? Some meals?
Bar space
Any other activities

*The Bathroom*

To be used only as a utility?
Door properly positioned with regard to fixtures, and privacy
Shower stall separate from tub
Tub not beneath a window
Nonskid surface in the tub
One or two basins (consider a cabinet beneath them)
Dressing table
Medicine cabinet, cosmetic space
Tissue and toilet paper dispensers practically placed?
Lighting
Extra electric outlets
Clothes hamper
Can linen storage be arranged in your bathroom?

*The Laundry*

Where do you want it (accessibility is important)—will constant stair-climbing be involved?
Washing machine and dryer
Tubs
Lighting
Extra electric outlets

# CHAPTER 12

# ALTERATIONS AND REMODELING

1. Why Alter or Remodel?
2. Plan Carefully Before You Start
   A. Remodeling an Old House
      1) The Old, Old House
      2) If You Want to Use a Contractor
      3) The Not-So-Old House
         a. The Interior
         b. The Exterior
         c. Checking the Zoning Regulations
3. Obtaining Estimates
4. Occupying a House During Alteration
5. Suggestions for Making It a Better House
   A. The Interior
      1) The Family Room
      2) The Living Room
      3) Bedrooms and Work Areas
      4) Attics and Basements
   B. The Exterior
6. Financing an Alteration

## 1. WHY ALTER OR REMODEL?

The perfect house is hard to find. Previous chapters on rehabilitating an old house or purchasing a lived-in house have warned that you must be prepared to face the remodeling of such houses. Even if you buy a nearly new house, you may want to make some changes in the kitchen, bath, or work areas or build an addition now or later. You may want to give an older house a complete facelifting. You must be very careful that such alteration or remodeling is not so expensive and inconvenient that it would be better to wait and purchase another house that meets your requirements more closely. Unfortunately, such a choice is rare. Most major alterations are done after a family has lived in a house awhile, and it is motivated by such things as additions to the family or the desire to modernize or enlarge because of increased income. Remodeling also may be done if you find an older house that meets your requirements for location, almost meets your criteria for livability, and the price plus the cost of remodeling is still less than the cost of a newer house, which meets these standards more closely. If, therefore, you are determined to alter and remodel, then you had better know how to go about it: how to pay for it, and how to deal with the contractor.

## 2. PLAN CAREFULLY BEFORE YOU START

This section is divided into several subsections because planning for the remodeling of a newly purchased old house is different from planning the addition of a room or altering the kitchen in the house you are living in.

ALTERATIONS AND REMODELING 139

## A. REMODELING AN OLD HOUSE

### 1. The Old, Old House

The question immediately arises: is it truly an old house or is it just a house which may be forty or so years old and is architecturally dated? Chapter 8 discusses the rehabilitation of the old, old house. It is assumed in the discussion that you plan to do a great deal of the work yourself. People who buy such houses usually do. It is also assumed that you cannot live in it comfortably until it is altered.

### 2. If You Want to Use a Contractor

If you want the work to be done by a contractor, then it is essential to get expert advice. If you cannot afford an architect, you probably know someone who knows a contractor, a carpenter, or a mason. First, go through the house yourself. Note everything that must be done, starting with the structural frame and then going through the other trades as outlined in Chapter 8. List in detail, with measurements if possible, what you think must be done. Then get your expert (or semi-expert) and go over it with him. You may have to pay for this advice. Have him translate your notes into more technical terms.

Next, find a contractor. If the house is in the country you usually can find a good small local contractor who does alteration work. Try to employ a general contractor in any case rather than a carpentry contractor, a plumbing contractor, a roofing contractor, and so forth. The general contractor will charge you some overhead for his time in supervising these subcontractors, and it is well worth it. If the house is in a large town or a city, ask for names of contractors at a local lumber yard or a general hardware store or look in the Yellow Pages. Get more than one contractor if you can. Go through the house with him. Make sure he understands exactly what he is to do. Have him give you an estimate which refers to your notes. <u>Do</u>

<u>not enter into an open-price contract</u>. If you can't get a firm bid you must set a cost limit—and get it in writing!

3. The Not-So-Old House

If you buy a house which is architecturally dated but otherwise sound and which meets your requirements for livability, you may not want to start any alteration work before you occupy it. This gives you an opportunity to study carefully the exterior architecture and the interior convenience before you start to plan or to spend money.

*a. The Interior* Again, note the things you would like changed. Such items as the layout of the kitchen work space, the sizes of the rooms, the location of a doorway, a bathroom with two entrances, or the need for more rooms or baths should all be studied. There is no reason why you cannot make drawings showing what you would like. Use an ordinary twelve-inch rule and make every quarter inch equal a foot. (This is known as "quarter-inch scale" in architectural drawings.)

It is recommended that you consult an architect after you have made this list. If you cannot afford an architect, you should try to have a knowledgeable person go over your list and your drawings. For instance, you should know whether you have enough electrical capacity for the new kitchen layout or the new room or wing; you should know whether your heating plant has enough capacity to handle the added space. If you add bathrooms, a dishwasher, or a washing machine, you may require a larger septic tank or drain field if you are not connected to a public sewage system.

When you have completed your list and have all the information you think you require, make a list of priorities. Very rarely can anyone afford as much house as he wants. You must, therefore, decide which items on your list you would like to have first. In making this priority list, it is wise to consider how long you think you will live in the house. Are you remodeling for children who will soon grow up and require other facilities? Is your job such that you may be transferred? Are you happy enough with the neighborhood and the house so

# ALTERATIONS AND REMODELING

that you will want to stay for at least ten years? You must realize that the changes you make and the money you spend are for your convenience only. Very rarely will the purchaser of your house pay the full value of these changes, because these alterations will not necessarily satisfy him.

*b. The Exterior* Although it is recommended that you employ an architect before you start any alteration or remodeling, it is not absolutely necessary for the interior work, because with a little expert help you probably can do well enough on your own. It is when you want to change the exterior that you may get into trouble. If you plan to add a room or two and a bath, or to finish the attic and add some dormers, you will want the lines and proportions of these additions to be in harmony with the lines and proportions of the rest of the house. Even if you think you have an eye for architectural style and good lines, you still will need someone who can make intelligible drawings so a contractor can give you an estimate. It is even better to make an arrangement with an architect. There are many architects and architectural draftsmen who work for others with no office of their own and whose charges are quite reasonable.

You may think that too much emphasis is being placed on the use of an architect. There are dozens of variations of Colonial, Georgian, and modern designs. There is an infinite variety of roof lines; there are shed or Dutch dormers, gable dormers, or flat-roofed dormers (sometimes called "doghouse" dormers). When you change the exterior you may have to install new gutters and leaders or to match a roof or windows. To define all of this properly takes a knowledge of construction as well as of architecture. If the specifications of the job are not set down properly, you are placing yourself at the mercy of a contractor.

*c. Check the Zoning Regulations* Either before or during the planning of any additions to your house you must check the local zoning regulations. At your local town hall or city building department you can obtain maps which show your location and zone. You should check the front yard, side yard, and rear yard requirements and the height limitations. If the addition which you are planning violates the limits set by the regulations, you will have to get a variance from the zoning authorities. These are often difficult to obtain. If you are buying a house to which you plan to add, it is a good idea for you to check the zoning before you buy. You may not want the house if you can't add to it. This restriction, of course, refers only to actual exterior additions and not to interior alteration or remodeling.

## 3. OBTAINING ESTIMATES

Earlier subsections of this chapter set forth the steps to be taken before you get in touch with a contractor. Of course, if you have an architect he will do this for you. If you are on your own you should try to obtain an estimate from more than one contractor. Have your list of priorities ready so that you can ask for separate bids on each item. For instance, you can ask for a price on adding a new bedroom and bath; another price for adding roof dormers; another price for enlarging the kitchen or an existing bath, and so on. Show the contractor through the house and show him your drawings and specifications.[1] Ask him for suggestions. A good contractor can save you money by suggesting alternate materials and shortcuts in construction. He can advise you about room sizes so that standard lengths of lumber or standard sizes of windows, doors, and frames can be used.

It is essential to get your estimates in writing and to make sure that the contractor knows exactly what you require. If the job is large it is recommended that you enter into a written contract. Standard forms of construction contracts are published by the American Institute of Architects. These forms

---

[1] A specification in architecture is an explanation in words of a drawing. In its ideal form it includes descriptions of workmanship and quality of material.

# ALTERATIONS AND REMODELING

contain the basic ingredients of a valid contract and *are fair to both the owner and the contractor.*

## 4. OCCUPYING A HOUSE DURING ALTERATION

Under the best conditions it is inconvenient and unpleasant to live in a house while there is any extensive amount of construction going on. If proper arrangements are not made, a house can become almost unlivable. Work which includes any exterior changes obviously should be done during the best weather when it is not too hot, cold, rainy, or snowy. Tearing off a wall of an air-conditioned house in a warm climate during a hot summer or of a heated house during a cold winter can create havoc. After choosing the best weather you can, take steps to isolate the areas where the work is going on by closing doors, where possible, and by tacking dropcloths or plastic sheets over all openings between the occupied quarters and the work. Arrange a schedule with the contractor so that he doesn't turn off your water on washday or leave piles of lumber or sand on your front lawn over the weekend. Always have him leave a clear path to either your front or back door. Be sure he protects everything he leaves overnight and weekends from the weather. If there is work that can be done indoors in bad weather and outdoors in good weather, you may be able to keep the contractor going steadily until he is finished. Indeed, one of your greatest problems will be to keep the contractor on your job without suffering long delays while he works somewhere else!

## 5. SUGGESTIONS FOR MAKING IT A BETTER HOUSE

There are many things that can be done—some of which are not too expensive—to transform an older house into a house conforming to contemporary life. Chapter 10 on architecture mentioned several of these features. The following sections will go into more detail.

### A. THE INTERIOR

1. The Family Room

One of the many conveniences in a single family house is the addition of a family room. This is really a revival of the older large kitchen where the family gathered to eat and where the children played and did their homework. Nowadays such a room is used for play, for television, for serving light meals, or for social gatherings. If possible, it should be near both the kitchen and the outdoors with a door leading directly to a porch or terrace. If you can, combine two smaller rooms or add an extension off the kitchen to make such a room. You will find it one of the most useful spaces in the house.

2. The Living Room

The living room need not be very large if you have a family room for informal entertaining. It always should be entered from the front door or a front hall. You can add bookshelves around the fireplace or change the mantel or add a large mirror over the mantel as a decorative touch. If it is necessary to do so, you could change a door location so that one does not have to walk through the room to go upstairs or to the rest of the living quarters. If there is no family room, then your living room should have space for your television and record player and for entertaining a sizable group of people. There is no ideal size for a living room—it should be large enough to suit your way of life.

3. Bedrooms and Work Areas

Chapter 10 discusses bedroom layouts and sizes for the most convenient living. Work areas such as kitchens, hobby areas and shops, and utility areas such as laundries, bathrooms, and fur-

# ALTERATIONS AND REMODELING 145

nace rooms are discussed in detail in Chapters 11 and 18. The suggestions and layouts mentioned there may be used whether you build a new house or remodel an old one.

4. Attics and Basements

The attic or basement of a house often can furnish an excellent opportunity for enlarging your living area with reasonable outlay. It is often possible to add an English or Dutch dormer to a Colonial type house and thereby provide space for one or two bedrooms and a bath (but this is a job for your contractor).

The basement can very easily provide space for a game room or a party room. This area should provide no difficulty for a reasonably skilled "do-it-yourselfer." You must provide ventilation if there are no windows or doors. This can be done by cutting an opening through the foundation wall and providing a fan which should be reversible for intake or exhaust. Panelling, wallboards, floor tiles, and ceiling materials are available at your local building-supply store. Be positively sure your basement is *dry* before finishing this area.

B. THE EXTERIOR

This section can be interpreted as a warning against too impulsive changes. Many people who buy an old house that has a rambling veranda or large bay windows assume that this dates the house and proceed immediately to remove these features. Please look carefully before you do this. Possibly a side porch can be turned into a delightful terrace with an awning and a low rail, and the bay window can add charm to your living room or dining room—a place to keep plants or to curl up with a book.

If you have to add a garage or carport, it should look like part of the house rather than an afterthought. If there is a detached garage see if it can be moved closer to the house and attached to it by a covered passage. Look at the houses around you and get the feel of the neighborhood before you start your exterior alterations. You don't have to do it all at once. It can be done in several carefully planned steps.

## 6. FINANCING AN ALTERATION

Alteration work can be financed in a number of ways. There is the bank loan which is specifically for this purpose and which bears a slightly lower interest rate than an ordinary bank loan. Such a loan can be paid off over a three-year period and one can obtain up to $5,000. In such cases the bank does not get a lien on your house in the form of a mortgage, it must depend on your personal credit. If you can qualify and if you are able to bear the burden of the comparatively quick payoff, this is an advantageous type of loan. Another way to obtain financing is through your mortgagee. When you arrange for a mortgage on a house, which you may wish to modernize, it is possible at the time of the "closing" to get a signed agreement from the mortgagee to advance funds over and above the amount of the original mortgage. Such funds are advanced as you complete the various steps of the alteration work. The advantage of this is that you obtain the same rate of interest on the advances as on the original mortgage. If you approach your mortgagee for funds without such an agreement, he may want to refinance (or "rollover") your mortgage at a higher rate of interest.

The Federal Housing Administration (F.H.A.) under Section 203(K) will insure a mortgage up to $12,000 for the purpose of altering, repairing, or modernizing a house. This amount plus the amount of the outstanding mortgage on your house must not exceed the total mortgage that your house would be eligible for under F.H.A. rules. You can inquire about this at your bank or building and loan society.

# CHAPTER 13

# BUILDING YOUR OWN HOUSE

1. Why Build?
2. Choosing your Building Lot
   A. Zoning
   B. Topography
   C. Buying Your Building Lot
3. Make Your Own Decisions First
   A. Decide What You Want
   B. Decide What You Can Afford
4. The Working Drawings
   A. Using Purchased Plans or the Builder's Designer
   B. Choosing an Architect
      1) The Architect's Contract and Duties
      2) Planning Meetings
      3) The Architect's Fee (How It Is Paid)
5. The Construction Contract
   A. Obtaining Bids
   B. The Contract
6. Inspection During Construction
   A. The Important Stages
   B. Construction Progress
7. Payments During Construction
8. Insurance During Construction
9. At Completion

## 1. WHY BUILD?

Why should you want to build your own house when there are so many completed houses available? It is a common belief that building a house causes the owner a great deal of trouble, that there is always a cost overrun, and that the contractor never finishes on time. These observations are not always true, but many people are frightened off and settle for buying someone else's house, which may not be quite what they want but is near enough.

The reason that you build your own house is that you want something which is uniquely yours, reflecting your mode of living and your taste. Building is an adventure—make no mistake about that. But it can be a very satisfactory adventure with a happy ending if you know what you are doing—and that is what this section is about.

## 2. CHOOSING YOUR BUILDING LOT

Chapters 2 and 3 of this book go into detail regarding your choice of a general locality, a neighborhood within the locality, and an actual building site. The suggestions and cautions in these chapters apply to choosing a site either for a new house to be built by you or for a house that you may wish to buy.

There are some differences, however, in your approach to the considerations involving the land if you are buying vacant land instead of a house.

### A. ZONING

Let us take zoning as an example. If you intend to buy an existing house it is important for you to know, first, whether

# BUILDING YOUR OWN HOUSE

the house complies with the existing zoning code, and second, whether you can add to it within the code or have to seek a zoning variance which is difficult to obtain. If you are purchasing a building lot, however, you have much more freedom. You can plan to build less house than the code allows and you can plan it for future additions within the code. On larger lots you have a number of choices as to where you can place your house and still remain within the code. You may even, if zoning permits, place your house so that you can sell off a piece of your land at a later time.

## B. TOPOGRAPHY

Here again you have more freedom to locate your house exactly as you want it with relation to sun direction, elevation, and slope of the land, and with subsequent additions such as a swimming pool, a terrace for entertaining or sunbathing, or facilities for other hobbies and activities. (Please refer to Chapter 3, Section 2D for more detail.)

## C. BUYING YOUR BUILDING LOT

When you have selected your site and have satisfied yourself that it meets all your physical requirements and suits your life style, then you may proceed to purchase it. One more suggestion before you make a final commitment: if you intend to retain an architect and know the kind of house you want, ask him to look over the site and to advise you as to whether you can build the house you want in that particular place. (See Section 4B in this chapter.)

The purchase of a building lot calls for payment in cash. Under normal circumstances mortgagees will not lend money on vacant land. It is advisable, therefore, to approach your local bank, insurance company, or building and loan society from whom you would like to borrow mortgage money to obtain their reaction to your proposed purchase. You should, if possible, obtain a commitment from a lender to the effect that if your prospective house meets his requirements he will lend you the money to build it.[1]

## 3. MAKE YOUR OWN DECISIONS FIRST

### A. DECIDE WHAT YOU WANT

Several previous chapters have described in detail how to decide on the number of rooms you want and how large they should be. Many rooms can serve multiple purposes if they are located properly in relation to other rooms; if they are of the right size and shape; if they have the proper light and are accessible to plumbing and heavy electrical lines. Decide on the exterior architecture—on whether you want a ranch house, a 1½-story Colonial, a two-story Georgian, or a completely modern house with a large glassed area, a flat or shed roof, and so forth. Decide also whether you want a basement; an unfinished attic; a one-, two-, or more-car garage, and how you plan to attach it to the house. How will your house fit its outdoor environment? (*see* Chapter 17 for details.)

At this time you can also draw single-line plans to scale of how the rooms will fit together. If you feel that you cannot afford an architect, you should look through magazines devoted to home building and even advertisements in newspapers which show plans of houses for sale. You should also go through the catalogs of the plan services to see if a house is illustrated there which meets your requirements.

---

[1]Chapter 15 on mortgages goes into some detail on this matter. If you want a conventional mortgage you must be prepared to pay most of the cost of the land as your equity. In FHA or VA-guaranteed mortgages you may be able to get by with paying only a small portion of the cost of the land.

## B. DECIDE WHAT YOU CAN AFFORD

It has been said before, and it bears repeating, that almost no one can afford as much house as he would like. It is necessary to go over your budget and examine your cash position and income (*see* Chapter 1) so that you may decide how large a house you can afford to build and to carry. Presently, mortgage interest is high and taxes are going up. You must take account of these factors. When you have decided on how much cash you can spend (allow 5 percent over for contingencies), and how large a monthly payment you can carry, you may then choose a house to fit these considerations. You can obtain an idea of the cost of local construction by asking a builder, your banker, a real estate broker, or the local tax assessor or building inspector.

Let us indicate how you would go about determining your cost.[2] You are told that local residential construction of reasonable quality and with no architectural tricks should cost $25 per square foot. A house of 1,000 square feet (25 × 40) should then cost $25,000. If you add a 200 square foot (20 × 10) garage at $10 per square foot you will add 10 × 20 × $10, or $2,000. Add your land cost of $5,000 (for a small building lot of 6000 square feet) plus some landscaping and a driveway ($1,000) and you come to a total of $33,000. (For a house this size you probably will not employ an architect.)

If you can borrow $25,000 it will cost you about $2,200 per year in interest and amortization and about $800 per year for taxes.[3] This means that your monthly payment for a 1,000 square foot house with a one-car garage on a 60 × 100 building lot will be $250. These figures vary widely from community to community, but for houses near large cities they are average. If you can afford $250 per month as rent (don't forget that a portion of this represents amortization on your mortgage and

---
[2] Please do not use these figures. They are only an example!
[3] You must determine this from your local tax rate (*see* Chapter 2, Section 2D)

can be counted as savings), you can build a 1,000 square foot house with a one-car garage. You can work this formula up or down, depending on your requirements.

### 4. THE WORKING DRAWINGS

After you have gone through all the steps to determine the size and price of the house you want, it is time to think about a building plan. The advantages of retaining an architect have been pointed out several times previously. A knowledgeable architect who will work with you in the design of your house to meet your needs, both expressed and implied, and who will be a shield between you and the builder, is worth every bit of his fee. A skillful architect often can save you a large part of his fee by holding down construction costs. The fact remains, however, that an architect's fee will amount to 7 to 9 percent of the cost of the construction, and to many people this is a critical amount of money.

#### A. USING PURCHASED PLANS OR THE BUILDER'S DESIGNER

Many local builders have architectural designers on their payroll who can modify a standard plan to fit your requirements. You can also purchase a complete set of plans and working drawings from one of several services.[4] Neither of these approaches offers you the cooperation of someone who is expert in construction and whose sole interest is to look out for

---

[4] One of these services is Reader Service Department, Better Homes and Gardens, P. O. Box 374, Des Moines, Iowa 50302. For one dollar they will send you a catalog of their plans.

your interest. You may be able to employ someone who is knowledgeable and who will charge less than an architect, but it is strongly recommended that you do not leave yourself entirely to the tender mercies of a contractor.

If you purchase plans you must also make sure that the plans and the specifications that come with them meet the local building and zoning codes. Furthermore, the placing of the house on the lot must also be in accordance with zoning regulations. You can obtain advice about this from a builder or from your local building inspector.

## B. CHOOSING AN ARCHITECT

If you decide to retain an architect because you feel that you can attain the fulfillment of your desires only through the personal service rendered by such a professional, then you must make sure that the architect you choose will be sympathetic to your ideas and will expand them. You also must mention at this time what construction costs you can afford. Ask the architects you visit to show you what they have done. It would be foolish for you to retain an architect who is closely identified with traditional Colonial architecture to design a flat-roofed modern house with large open areas and large glass areas, and vice versa.

Tell him what you have in mind and obtain a reaction. He should show an understanding of what you are trying to accomplish. A good architect should give you more than just working drawings. You can buy those for $50 or so.

### 1. The Architect's Contract and His Duties

The American Institute of Architects publishes a printed form of contract between owner and architect which sets forth in detail the duties of the architect and how he is to be paid; as in any contract between parties, it also mentions the responsibilities of the owner (you).

For your quick understanding, a brief summary of the highlights of an agreement between architect and owner follows:

1) The architect should prepare one or more schematic line drawings for your approval to show whether he has a full grasp of what you require.

2) The architect should at this time give you an estimate of what the construction cost of the various schemes will be. Many architects ask the opinion of a local builder. These estimates serve to warn you and allow you to modify your requirements before any expensive working drawings are started. In the unlikely event that the estimated cost is lower than anticipated, you can add to your plans.

3) In the next step the architect prepares further drawings which progress from the schematic line drawings to show the structure in some detail as to room size, layout, and a certain amount of detail regarding the heating, plumbing, and electrical systems.

4) At this point the architect should again check with a builder to reestimate the cost of construction.

5) If these preliminary scale drawings meet with your approval and the estimate is within your budget, you can have the architect proceed with complete working drawings and specifications.

6) When the plans and specifications are completed the architect will help you select a list of bidders and will assist you in evaluating the bids and in choosing the contractor. At this point he should be sure that the contractor has included everything necessary for a completed house. If anything is left out you may be plagued by unforeseen extra costs.

7) When you have awarded the contract, the architect must then supervise the construction to be sure that the contractor is furnishing materials and labor in strict accordance with the plans and specifications. In order to do this properly he must do the following:

    a) Visit the job at frequent intervals so that he can inspect the work at crucial stages of the construction (foundations, framing, plumbing, electrical and heating systems, finish trades).

# BUILDING YOUR OWN HOUSE

 b) Act as the interpreter of the plans and specifications in the event of a dispute between the owner and the contractor.

 c) Certify to the owner and the mortgagee that the contractor is entitled to progress payments as requested.[5]

2. Planning Meetings

The preceding section has outlined the duties of an architect from conception to the completion of your house. It is well, however, to go into considerably more detail regarding the first step that you and your architect take together. This is the step during which the schematic drawings are prepared.

It should be emphasized that it is during the preparation of the schematic drawings that your dream house begins to become a reality. You should examine these drawings carefully and again be sure that the architect knows exactly what you want, and it is at these planning meetings—there should be several of them—that you come to an agreement as to how much house you can have. It is part of the architect's duty to be constantly aware of the construction cost of the things you want and to warn you if your wants are exceeding your ability to pay for them. The author has seen many cases in both residential and business construction where the contractor's estimates so far exceeded the budget that the owner and the architect had to go through the wearisome and melancholy task of cutting down dimensions and even cutting out rooms. In such a process it can be very easy to lose the whole spirit of the house. Please be careful about your requirements and take heed of what your architect advises.

It is at these planning meetings that the architect or whoever advises you should be aware of the foundation condition at your site. The largest single unknown factor in any construction project is what lies beneath the surface. In Chapter 3, Section 2D your attention was called to the possibility of building on

---

[5]A schedule of typical progress payments wil follow in Section 7 of this chapter.

filled land or on a former marsh; there is also the possibility of finding ledge rock under part of your foundation or of finding ground water only a few feet down. In such circumstances you may have to waterproof your basement or blast ledge rock or drive pile foundations. Where there is any suspicion that any of these conditions may be present (you can ask near neighbors or the building inspector or the prospective bidders), you would be well advised to have one or two test pits dug to see what the soil looks like underneath and whether the pit fills up with water when there has been no recent rain.

If you find such conditions or a strong likelihood of them, you must ask the contractors to include in their bids as a separate item the cost of performing the additional work required to provide you with a secure foundation and a dry basement.

3. The Architect's Fee (How It Is Paid)

Architect's fees vary from state to state and usually are set by a statewide professional architectural society. In the form of agreement which is most often used for residential construction, the fee is based on the total cost. This is known as the percentage of construction cost formula and the total costs are arrived at in negotiation with the contractor.

The fee arrangement provides, however, that if you make drastic changes after you have approved the plans or if you require the architect to perform additional services such as interior decorating or preparing large change orders, you must pay additional fees, usually based on an hourly rate.

The architect's fee is paid in stages. You should make an arrangement with him to pay him for only what he has done to a certain stage, so that you will be protected if you decide for any number of reasons to abandon the project. This can be done as follows:

1) You come to an agreement regarding his total fee, which is based on a percentage of the total estimated cost.

2) You then arrange to pay him 15 percent of this fee if you abandon the job at the schematic stage or 35 percent if you

allow him to go on to develop the schematic drawings to the design development stage which shows your house laid out to scale with the utilities defined.

3) If you order him to continue and complete working drawings, to obtain bids, or to help you obtain bids and then to help with the negotiations, you will have to pay 80 percent of the fee.

4) The last 20 percent is paid for construction inspection.

The architect's contract form mentioned previously in this chapter tabulates these payment stages, but it is hoped that this section will further clarify the language.

## 5. THE CONSTRUCTION CONTRACT

When the plans and specifications have been completed by your architect or when you have purchased plans and have had them revised to your satisfaction, you are ready to obtain estimates from several contractors.

### A. OBTAINING BIDS

Either you or your architect must determine the qualifications of the contractors from whom you intend to solicit such bids. If you do not have an architect, then you must investigate. Find out who the good builders are from the local building materials dealer, the building inspector, your bank, or a real estate broker. If you can, you should look at some houses they have built, and if the owners will tell you, find out whether the contractor met his time schedule and his estimated cost. Inquire particularly if he requested extra money even for small changes during construction.

If you do it yourself you should see each contractor and

obtain an impression of his understanding of your requirements. Invite all of them to call you (or your architect) if they have any questions regarding the meaning or intent of any part of the plans or specifications. Ask as many times as necessary if the contractor has enough information to produce a completed house ready for you to move into.

When the bids are received you should look for several important items:

1) The site and the proposed building should be described (for example: "On Lot No. 123 on Front Street between First and Second Avenues we propose to erect a seven-room house with full basement and with attached two-car garage").

2) The plans and specifications on which the bid is based should be described by date and number.

3) The sum of money to be paid the contractor must be stated clearly with no strings unless the request for bid has asked for prices for alternate items—for example: "Please quote additional cost if the entire house is air-conditioned in accordance with details shown on Sheet No. X"; "Please quote additional price if pile foundations or spread footings are required as shown on Sheet Y"; "Please quote additional price for waterproofing basement as described in Sections XY of Specifications."

4) The bid should state how much time you have to accept it.

5) The bid must state how much time the contractor requires to complete the house *ready for occupancy*. (This means until he obtains a certificate of occupancy from the local authorities.)

6) The bid should state, or you must find out, who carries what insurance during construction.

If the bids are too high or are in any other way unsatisfactory, you have the right to refuse to accept any or all of them. You also can see the lowest bidder or any of the bidders and attempt to negotiate a lower price with him or them. This is where a skillful architect or building expert can be of help. In the negotiation the contractor can be asked for his suggestions on reducing the cost by omitting non-essential items, by substituting less expensive material, or even by changing some architectural details. He should quote a price for each change and you can then decide which things you must have, which you can add later, and which you can do without. At this point you are ready to enter into a construction contract.

## B. THE CONTRACT

As in the case of the architect's contract, there is a standard form published by the American Institute of Architects which describes the duties and obligations of the architect, the owner, and the contractor. This contract should contain the same important items that are mentioned in the bid (discussed in the preceding subsection). The standard form also mentions the following salient points:

1) The contractor must use his best efforts to produce a structure that will conform in all respects with the plans and specifications and any other written understandings (called the Contract Documents).

2) The contractor is to pay for all the cost of the work including labor, material, permits, license fees, and any other fees or permits required by various public authorities.

3) The contractor is responsible for compliance with all codes, ordinances, rules, and regulations.

4) The contractor must furnish samples of materials as requested by the owner or the architect for their approval. He also must furnish shop drawings as called for by the architect.

5) The contractor is responsible for safety on the premises.

6) The contractor is responsible for certain insurance. (*see* Section 8 of this chapter.)

The architect's duties during construction have been mentioned previously, but an additional point of considerable importance must be made here. In both the architect's and the contractor's contracts there is a saving clause which to some extent relieves the architect of responsibility.[6] This clause states that the architect will visit the job periodically "and will endeavor to guard the Owner against defects and deficiencies in the work," *but it also states that "The Architect will not be required* to make exhaustive or continuous on-site inspections to check the quality or quantity of the Work." These statements may seem to be contradictory, and you, the owner, are advised to have a clear understanding with your architect as to how often he will visit the job and how carefully he will check it.

## 6. INSPECTION DURING CONSTRUCTION

Proper inspection of your house while it is being built is most important, to be sure that you get what you are paying for and to discover any inherent defects in the construction which may cause trouble or even structural failure later. This is not to say that the builder is dishonest or trying to get by with inferior work. It is simply a fact of life that workmen make errors and that in an honest effort to speed the job the contractor will substitute a "just as good" material for the one specified.

The author, when he built his present house over twenty years ago, allowed the builder to substitute "just as good" window frames and sashes because the ones specified would not be available when required. The delay might have been a week or two. The frames and sashes which were used looked just as good but were not. They have been a small nuisance for twenty years—a high price to pay for saving a week or two during construction.

---

[6]A saving clause is one which creates an exception to the spirit of a contract or an understanding.

# BUILDING YOUR OWN HOUSE

## A. THE IMPORTANT STAGES

In Chapter 5 ("The Development House") there are descriptions and drawings of proper foundations, structural framing, and roofing as well as sections of a typical building code and descriptions of what constitutes good interior finishing. These descriptions and cautions are meant more for those who are interested in buying in a development than for persons who are having a house built for them. Inasmuch as the contractor who builds to order generally has a higher standard of workmanship and presumably is building in accordance with your plans and specifications, he is not so apt to cut corners.

Nevertheless, you must make certain that your architect or some other competent person inspects the construction at frequent enough intervals so that every part of the construction can be seen before it is covered. Such inspections are vitally important at certain stages of the construction, which are described in the following paragraphs.

1) *When the excavation has been completed for the wall footings and the contractor is ready to pour the concrete for these footings.*

At this point most building codes call for the building inspector to examine the bottom of the excavation to make certain that it is firm, undisturbed soil or rock. This bottom takes the entire weight of your house. Either your architect or an expert employed by you should also inspect it. If the building site is on filled land or on the site of a former marsh, wetland, or swamp, be particularly vigilant. Severe settling of foundations can cause extensive damage. A conscientious building inspector may insist that the contractor drive piles for the foundations when the underlying soil is unstable. In less severe cases, spread foundation footings may be enough.

If you have been alert to the possibility of such a condition you will have allowed for it when you obtained bids.

2) *When the foundation walls have been completed.*

If the foundation walls are of concrete they will be less

trouble than those of concrete block. For at least the top layer, the concrete block should be solidly filled with mortar. This serves as a solid foundation for the wood sill that is supported by it; as a solid barrier against termites which may come up through the hollow block; and as a means of securing the bolts which join the wood sills to the foundation.

At the foundation stage you also should inspect the sewer line to see that it is properly pitched—*at least* a three-inch downward slope toward the sewer line in the street (gravity feed). And make sure there are not tree roots near it; growing roots can cause havoc with a sewer line.

3) *When the framing is being completed or, in the case of a brick or block house, as the walls are being built.*

In the first case the wood frame should be securely nailed, the floor joists should rest solidly on sills and girders, the roof beams or rafters should rest solidly on the supporting walls, and the entire structure should be rigid. In the case of brick or masonry walls, you should look for solid mortar joints on both sides of the wall. If a cavity wall (*see* Figure 16) is called for you must make sure that there is a clear space between the two portions of the wall. Poorly laid masonry can result in damp exterior walls.

4) *When the roof is being laid and the door and window frames are being installed.*

There should be copper or aluminum flashing over every door and window frame; around every opening in the roof for the chimney, plumbing stacks, and so forth; and at every point where the roof line changes direction (as for dormer windows) or where there is a valley. The flashing serves as a continuous shield against the weather. *Well-installed flashing is your best protection against leaks through walls or roof. (see* Figure 6.)

5) *As the plumbing, heating, and electrical work is being installed.*

There is not much that an amateur can do about checking proper workmanship or materials in these trades. You will have to rely on the builder if you don't have an architect or other

FIGURE 16

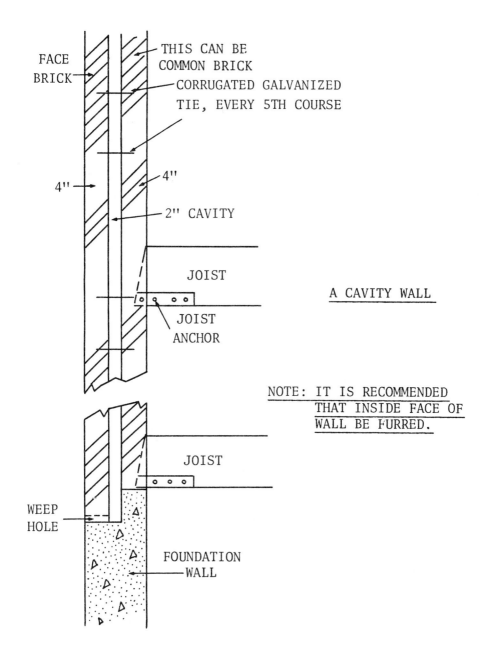

expert checking the work. Studying the codes to see that your water and waste lines are of the right size and that your electrical panel board has sufficient spare circuits can be interesting, but it also can be time consuming. During installation of heating, plumbing, or electricity you also should be alert to the fact that the men installing these lines are likely to saw or chop away large hunks of floor joists or wall studs to make their job easier. Electrical work should not be started until the house is weathertight.

6) *As the finish work is being done.*

The finishing trades in construction are the ones which cover all the rough framing and piping and electrical cables and produce a house ready to live in. They include finished floors, cabinet work, plaster or wallboard, tiling, hardware, and painting. Unlike mechanical, electrical, or structural work, the finishing of a house is completely visible and is what you will always see. You are as good a judge as anyone of what constitutes good work. Here are some of the things to look for:

a) *Finished Floors:* If the floors are wood, look for wood with no knots, tight joints, and secure nailing. If they are asphalt or vinyl asbestos (the latter is much better) laid over concrete, look for straight lines, tight joints, and full tiles.

b) *Plastering or Wallboard:* Plastering over metal or gypsum lath is an almost lost art in small house construction. Even quite expensive houses now use gypsum wallboard which comes in sheets four feet by eight feet. The board is *securely* nailed to wood studs and the joints are filled with a plasterlike material into which a perforated strip of cloth tape is pressed. The plaster must then be feathered out (this should be done in three operations) so that no joints will be visible when the walls are painted or papered. The wallboard should be at least 5/8 inches thick.

c) *Cabinet Work and Finished Carpentry:* Look for tight joints in the moldings. Slide windows up and down and open and shut doors to see that they hang level and click shut smoothly. The installation of hardware is done by a carpenter. Note whether it fits properly. In this connection also you must

# BUILDING YOUR OWN HOUSE

insist on exterior door locks with positive bolts that cannot be pushed aside by an intruder. This also holds for window latches.

d) *Painting:* Be sure you get the number of coats that you are paying for and that the paint comes in sealed cans and is the brand specified. If the brand is not named in the specifications, ask about paint quality at a good local paint store. Look for drips or runs and unpainted inside corners.

When you or your architect or expert—or all of you—have carefully and intelligently looked at your house under construction, and if you have employed a reliable contractor as well, you should have a well-built, weathertight, well-finished house that will be a source of pleasure to you for many years.

## B. CONSTRUCTION PROGRESS

The construction of an average-sized single-family house (two to four bedrooms) should not require more than four months from start of excavation to completion. Because of bad weather or the possibility of unforeseen delays, you may have to add a month or even two. It is unfortunate but true, however, that very often the rate of progress depends on the constant alertness of the owner or architect in following the job. The majority of small builders simply do not have the time, energy, or facilities to set up a proper progress schedule and to follow up on it.

This section will give you some guidelines which you can follow and discuss with the contractor. Some of these may seem so simple that you will wonder why it is necessary to mention them, but it is sometimes surprising how many seemingly obvious things can be forgotten or overlooked in the rush and excitement that often goes with building a house.

1) Before the excavation starts, the contractor must make arrangements for electrical power and for connecting to street water mains and sewer lines.

2) While the excavation is under way, the contractor should

prepare a clear path for concrete trucks; if the foundation walls are to be cement block, he should have them on the job.

3) When the foundation walls are under construction he should have the framing material delivered. If it is to be a brick or block house, this material plus floor joists, whether of wood or light steel, should be on the job. If it is a wood frame house, the sills and studs and girders and joists should be on the job and piled in the order of their use. If it is a modular or prefabricated house, the essential framing sections should be on the site.

4) At this point or even at the very start of the work you should ask whether all the possibly hard-to-get material has been ordered and what the delivery dates are. Such material can be special tile, colored bathroom fixtures, radiators, furnaces, fans or other parts for air conditioning, special doors or hardware, or even such simple items as clear oak flooring.

5) When the framing is under way the contractor must prepare for the delivery of sheathing and roofing material because the house must be made as weathertight as possible as soon as possible. He also should have a mason on the job for the chimney, fireplace, and any other outside masonry.

6) The plumber and heating contractor start as soon as the framing is completed and the house is reasonably weathertight. The electrician cannot start until the roof is on and the windows and doors are in place. The builder, therefore, must arrange to have these subcontractors on the job as soon as the framing and sheathing are complete.

7) When the house is weathertight, the contractor should start his wallboard men. These men cannot start until the electrician, the plumber, and the heating men have completed their roughing. If the house is being built during the cold weather, arrangements should be made for temporary heat so that the interior finishing trades can perform their work. It is sometimes possible to start the permanent heating system at this point, and doing so is strongly recommended as temporary heating devices can be troublesome.

8) At this time the finished floors should be started and followed by cabinet work, doors, and trim. The tile man also

should be working; when he is finished the plumber can install the bathroom fixtures.

If you have read the foregoing carefully you will see that the construction of a house follows a logical sequence. It requires only some forethought and vigilance for it to be completed within a fixed schedule.

## 7. PAYMENTS DURING CONSTRUCTION

All mortgagees set certain stages of construction at which they will make payments to the contractor. There are usually four or five points in the progress when such a payment is made. These points are approximately as follows:

1) At the completion of the excavation and foundation work and the beginning of the framing.

2) At the completion of the framing and the start of rough plumbing and heating work.

3) When the structure is enclosed and all the mechanical and electrical roughing is complete.

4) When lathing and plastering are completed and finish carpentry and cabinet work are under way.

5) At full completion when a certificate of occupancy is issued by the building inspector.

Any of these stages can be combined in a number of ways to make a four-payment plan. The mortgagee will always send someone to check the progress and will require a release of lien from the contractor before a payment is made.[7] Very often they will require the approval of the payment by your architect if you have one; this is one more safeguard for you.

---

[7]The release of lien protects the mortgagee and you against future claims for non-payment from the contractor or any of his suppliers or subcontractors.

## 8. INSURANCE DURING CONSTRUCTION

Chapter 14 will discuss the insurance on a completed house. It is the purpose of this section to discuss only the insurance that is necessary while you are building your house and before your permanent insurance takes over.

You can be in a vulnerable position while your house is under construction and you must cover your risk very carefully. First, you should have the successful bidder produce certificates of workmen's compensation, public liability, automobile liability, and property damage in amounts which are large enough to cover present-day awards. He must protect you against all claims by workmen employed by him or his subcontractors or by the general public. Coverages such as $150,000 per person, $500,000 per accident, and $50,000 property damage are quite common. Workmen's Compensation limits are set by state law. Second, you as the owner must cover the premises against fire and other perils such as windstorm, vandalism, and malicious mischief. This insurance covers the contractor, his subcontractors, and material on the site. It is recommended that you get the completed value form which covers the final completed cost of construction right from the beginning.

Be sure to have your insurance broker advise you about this and have him examine the contractor's certificates as well.

## 9. AT COMPLETION

Contractors always seem strangely reluctant to complete a house. When you think the house is almost ready for you to move into, start requesting a certificate of occupancy which the contractor must obtain. Go through the house and look for any lapses in good workmanship. Then, move in. This is the act that finally pushes the contractor out. Don't approve the final payment until all the odds and ends are completed. It is difficult to get a fully paid contractor to come back.

Good luck and enjoy your house!

# CHAPTER 14

# INSURANCE

1. The Basic Facts
2. Insuring Your Home
3. The Available Policies
4. Other Insurance

## 1. THE BASIC FACTS

It certainly should not be necessary to urge people to carry insurance on their homes, their furniture, their automobiles, and their lives. Yet many people are strangely reluctant to carry the amount and kinds necessary for full protection against unforeseen hazards.

It is true that insurance premiums have risen and that paying for full insurance coverage may be a hardship. But consider the fact that <u>a disastrous fire, windstorm, or car accident can wipe out a lifetime of savings.</u> Although you may feel that your insurance payments are high, do not forget that you individually never pay the full amount of your own losses. Your insurance premium goes into a "kitty," or general fund, together with the premiums of millions of other people, and the claims are paid from this premium fund. What is more, you are protected against paying unreasonably high premiums because state governments carefully regulate the insurance industry and no premium rates can be set without their approval. Let us hope that you are one of the lucky ones who will never have to collect a major claim. Be content just to pay and sleep well! One need only read the headlines of jury-rendered judgments to be convinced that adequate insurance is a necessary fact of life.

## 2. INSURING YOUR HOME

If you have a mortgage on your home the mortgagee will insist that you carry enough insurance to fully protect *his* interest. This section will inform you about protecting *your* interest as well. The most important coverage for a homeowner is fire insurance. With today's rapidly rising construction costs, you must be sure that your house is insured for its full replacement value. Do not forget that any loss over the stated insur-

INSURANCE 173

ance amount comes out of your own pocket. The usual homeowner's policy includes an additional amount (usually 50 percent of the dwelling value) applicable to personal possessions, and if your house is underinsured you may also suffer loss in this regard.

## 3. THE AVAILABLE POLICIES

This chapter and this section about policies do not pretend to tell you all about insurance. They simply call your attention to the available kinds of insurance with a brief description of the coverage. *You should carefully note what you are and are not insured against.* Ask your broker about the exclusions. For example, you could be in serious trouble if you think you are insured against damage caused by high tides, surface or flood water, or frozen pipes—and then find out you are not.

There are several forms of homeowner's policies. The simplest form is for *Fire and Extended Coverage* (sometimes called HO-1 or HO-A) and provides insurance on home and contents against fire, lightning, windstorm, hail, explosion, riot, civil commotion, vehicles, aircraft, smoke, vandalism, malicious mischief, breakage of glass, and theft. This seems like a fairly complete list of risks—and it is. However, a close reading of the "fine print" in the policy will disclose that such risks are covered only within fairly narrow limits. For instance, explosion covers an explosion only if it occurs in the combustion chamber or connecting flues of your furnace, but steam boilers are excluded. Smoke insurance covers only smoke caused by a heating or cooking unit within the insured premises but not from a fireplace or other sources. Vandalism or glass breakage is not covered if you leave your house vacant for more than thirty days immediately preceding the date of loss.

Form B or HO-2 provides a wider coverage and you should look into this carefully. But the best homeowner's policy to get—if you can afford it and don't wish to take the risk of being self-insured[1]—is the *Homeowner's Special Form* (HO-3 or HO-C). This policy covers you against all the risks in the HO-1 form, plus such others as damage to building or contents from falling objects (trees falling on your house, from wind or weight of ice or snow); collapse of buildings; sudden and accidental tearing asunder of steam or hot water systems or of appliances for heating water; leakage; and freezing of plumbing, heating, and air-conditioning systems.[2]

All homeowner policies allow you additional money in case of a major disaster—for instance, if you had to live away from home if your house became uninhabitable—and also cover you against personal injury liability to persons other than employees who enter your property (your dog biting the mailman or someone slipping on your front steps).

You can further extend your insurance coverage to objets d'art, books, jewelry, furs, fine furniture, cameras, stamp or coin collections, and other scheduled personal articles if you list such objects and obtain a certified appraisal from an expert acceptable to your insurance carrier or if you present proof of their cost. Although this chapter has outlined the kind of insurance that is available, you are strongly urged to obtain the advice of a reputable insurance broker licensed by the state.

## 4. OTHER INSURANCE

Although this book is about home owning, it is well to mention here other insurance because inadequate coverage of other

---

[1] Which means you pay for all losses or damages out of your own pocket.

[2] Although the Homeowner's Special Form HO-3 is far broader in coverage, there are still exclusions applicable such as flood, surface water, tidal water, seepage, earthquakes, and so forth. Close attention should be paid to such restrictions.

# INSURANCE

risks may cause you severe financial hardship—and even can place your home ownership in peril. The most important of these is automobile insurance, and the most important kind of automobile insurance is liability for bodily injury and for damage to the property of others. Requirements vary considerably in different states. Some do not require any automobile insurance until *after* the first accident. In others, insurance is compulsory *before* the car may be licensed. Many states have so-called Financial Responsibility laws specifying minimum liability limits which must be carried. In view of the high judgments being handed down by juries these days, it is foolhardy for a homeowner to risk losing all he has worked for for many years, or even possible bankruptcy, just because he failed to carry higher limits than the minimum. Depending on your financial circumstances, of course, a minimum of $100,000 per person, $300,000 per accident, and $25,000 of property damage is recommended. It is not at all uncommon to carry $500,000, $1,000,000 and $50,000. The cost of such policies is surprisingly low when compared to the cost of the basic policy. Along with these basic coverages, you also have a choice of other coverages such as medical payments, fire and theft, collision (which covers damage to your own car), and other risks. Investigate all of them and decide which premiums you can afford and for which you are willing to become self-insured. You should also be aware of the fact that automobile insurance is quite difficult to obtain now, especially for the young and for senior citizens. Do not lightly give up any policy you may now be carrying—whether it be a homeowner's policy, an automobile policy, or any other type of coverage.

Worth noting, among the many types of life and income insurance available for the homeowner, are mortgages which carry life insurance as part of them. Under such policies the house will be paid off automatically by the insurance company in the event of death of the breadwinner, (leaving wife and children with a home free and clear).

CHAPTER 15

MORTGAGES

1. What a Mortgage Is and What It Does
2. The Kinds of Mortgages That Are Available
    A. The Federal Housing Administration (F.H.A.)
    B. The Veterans Administration (V.A.)
    C. Conventional Mortgages
3. Closing Costs
4. Title Insurance
5. Truth in Lending

## 1. WHAT A MORTGAGE IS AND WHAT IT DOES

After you have determined approximately where you would like your house to be, what kind of house it should be, how large it should be, and how much money you have available for the down payment and upkeep, the next step is to determine how much money and on what terms you can borrow to help pay for it. In return for lending you the money, the bank or other lending institution will receive a mortgage on your house. A mortgage is the legal paper by which you pledge your house to the lender as a security for the payment of the debt. If you do not make payments in the amount and at the time stipulated by the mortgage, the lender can, subject to local laws, obtain possession of your house. This is no different from the penalties which a landlord can invoke for nonpayment of rent. The difference is that in the foreclosure of a mortgage for nonpayment your cash interest in the house is almost always wiped out. This is another reason for carefully assessing your financial situation before you venture into the purchase of a house. Millions of homeowners have mortgages. Just be careful.

Houses usually are not purchased for cash because it is not economically sound to do so. Cash is a marketable commodity and normally should not be tied up in anything as static as a house. A supply of available cash is a great comfort when there is financial trouble but it is not always easy to obtain a mortgage at a moment's notice. It is much better to obtain such a mortgage when you buy or build your house. You should familiarize yourself with the mortgage market. The amount of money which a financial institution is willing to lend on a house is very often an excellent indication of the house's value. A house with a mortgage on it is much easier to sell than one for all cash.

The present method of paying off a mortgage is a painless way of saving. Each monthly payment is the sum of two items—the interest on the outstanding balance of the mortgage and the *amortization* payment which reduces the amount of the mortgage each month. Each monthly payment throughout the term of the mortgage is equal. During the first part of the term the interest payments are higher than the amortization payments. As the mortgage term progresses, however, the principal amount becomes lower because it is being paid off by the amortization payments and therefore the interest gets lower; because all payments are equal, the amortization becomes higher, and this is the money *you are saving*.

## 2. THE KINDS OF MORTGAGES THAT ARE AVAILABLE

Mortgage money on single-family houses is normally available from many sources. Of late there has been a money shortage, and with the high rate of interest available on bonds and other investments, the mortgage lending institutions were inclined to place their available funds in these high-paying investments and to avoid the private house mortgage market. Another reason for the shortage of mortgage money was the usury laws of many states which forbade the charging of interest over a certain percentage to a private person. Recently, however, the states have amended their laws to allow higher interest and some banks and other lending organizations have bettered their interest rates by making service charges or discounting a loan in advance (called "charging points"). In other words, you actually may obtain only $9,800 on a $10,000 face value mortgage but you will be paying interest on the entire $10,000. This tight money situation made it difficult for the average home buyer to finance his purchase, but times are changing and it is getting progressively easier to obtain mortgage money.

### A. THE FEDERAL HOUSING ADMINISTRATION (F.H.A.)

This federal agency was established in 1934 during the depression. Its purpose is to facilitate the granting of mortgage money by private investors, based upon a government guarantee to take over the mortgage in case of a default; and to issue interest-bearing debentures to the investor in the amount of the value of the mortgage in the event of such a default. This in effect backs the credit of the homeowner with the credit of the government. Under the protection offered by this guarantee, the F.H.A. effectively standardized the rules and terms for mortgage lending and upgrading private housing construction.

Under F.H.A. rules the guaranteed loan may be made only by an approved mortgagee, which further protects the prospective homeowner against sharp practice. Its rules also provide for uniform appraisals, for a standard form of application for a mortgage, and for certain requirements which the borrower must meet. These requirements are a good credit rating; the possession of the cash required at the closing of the mortgage; and a steady income sufficient to make the monthly mortgage payments without difficulty.

In cases where a development house is being purchased or a house is being built for sale, the buyer can obtain some assurance that the house will meet minimum standards of construction if the mortgage is insured by F.H.A. Where building has not yet started when the mortgage commitment is made, F.H.A. will insist that a one-year warranty be given by the builder that the house will meet the requirements of the plans and specifications on which the mortgage commitment was based. This is an added protection for the home buyer.

An F.H.A.-insured mortgage can be obtained on a single-family house up to $33,000, and the actual amount is determined by an appraisal of the property by the agency. The ratios of amount of loan to appraised value vary between houses being built, houses newly completed, and older houses. On a house priced at $25,000 and appraised at that figure, the F.H.A. may guarantee a loan from $21,250 to $23,550. The down payment will be the difference between the price of the house and the amount of the mortgage plus the closing costs (which will be explained later).

An F.H.A.-insured mortgage can also be obtained on a mobile

home; a cooperative which has been released from a project blanket mortgage; certain types of condominiums; single-family homes in urban renewal areas (for example, an old brownstone in a central city area); and several other types of housing. The prospective house buyer can obtain information about the F.H.A. rules and regulations and their rates of interest and other costs from his local banker or building and loan society, if they are approved lenders; or by writing to the U.S. Department of Housing and Urban Development (HUD) in Washington, D.C.

Interest rates charged for F.H.A.-insured loans are competitive and vary with the money market. In recent times these rates have varied by as much as 5 percent. The monthly payments required depend, of course, on the interest rate that is in effect at the time of the closing of the mortgage.

The F.H.A.-insured mortgage and the V.A.-insured and -guaranteed mortgage (on which details follow) are excellent means whereby a prospective home buyer without much cash can buy a house and can to some extent be assured that an unscrupulous builder or developer is not taking advantage of him. They are by no means the only way of financing a house, and in the long run no government or other agency, no matter how well intentioned, can substitute for the house buyer who knows what he is doing.

B. THE VETERANS ADMINISTRATION (V.A.)

Under an act of Congress of June 22, 1944, the Veterans Administration was authorized to partially guarantee or insure loans made to qualified veterans by public or private lenders and to make direct loans in areas where other money was not available. These "G.I." loans can be made for other than single-family homes, but this section will confine itself to this portion of the laws. To be eligible for such a loan under the original law, a veteran had to meet certain requirements regarding the length of his active duty in the armed forces and the time between his separation from the armed forces and the date of his application for a loan; and he must have had an honorable discharge.

The V.A. program has been amended from time to time to allow for higher interest rates. The last amendment was signed by the President on October 23, 1970, and is called the Veterans' Housing Act of 1970. Under this act a veteran's right to obtain a G.I. loan guarantee on his house has been extended for his lifetime, subject to any previous commitments he may have with the V.A. and an honorable discharge. It also allows a higher guarantee on such loans and it extends the protection of the loan guarantees to mobile homes and to condominiums. In many instances a veteran may be able to purchase a home with no down payment. The qualified veteran who wishes to make a G.I. loan must go through the following steps:

1) He must visit or write to the nearest regional office of the V.A. in order to obtain a Certificate of Eligibility. If he has met all the requirements of length of service and has an honorable discharge, the V.A. will issue a certificate which sets forth the amount of the loan guarantee and the terms and conditions under which it is made.

2) When he has found the house he wants, the veteran can take his Certificate of Eligibility to any bank or lender; if they are willing to make the loan they will ask the V.A. to appraise the house.

3) The V.A. will then have one of its fee appraisers examine the property and, based on his appraisal, it will issue a Certificate of Reasonable Value which tells the lender how large a loan he may make which will be fully guaranteed. If the lender so requires, the V.A. also will issue a Certificate of Commitment.

# MORTGAGES 183

        4) When such certificates have been issued, the veteran may obtain his loan and purchase his home if he also meets certain minimum credit requirements such as a steady job and some money in the bank.

There are actions the V.A. can take to help the veteran in the event of default, but for the purpose of this book let us not foresee any such unhappy event. It must be remembered, however, by the veteran, that the G.I. loan is a *loan*. It is *not* a gift and *it must be repaid*.

## C. CONVENTIONAL MORTGAGES

A conventional mortgage is a mortgage the house purchaser can obtain from an insurance company, a bank, a building and loan society, or any other lender who does not rely on guarantees from any governmental agency—but who does rely on the credit and reputation of the borrower and the continuing value of the property on which the loan is made.

Commercial banks generally do not make such loans, but savings banks do and building and loan societies consider the single-family home a major source of their investment business. The prospective house buyer who wishes to obtain a mortgage loan is advised to approach his own bank or his own insurance company first. Even if it cannot make the loan because the property is out of its territorial limits or because of shortage of funds or other reasons, the mortgage officer of the institution can give valuable advice about rates and may recommend another lender. At this time it is presumed that the applicant for the loan has located the area where he wants to buy or build and has made up his mind as to how much he wishes to spend and the kind of house he wants.

When the final choice of property has been made—at this time it may be only a piece of vacant land and a set of house plans—the prospective purchaser should approach the lender of his choice and make formal application for the loan. Unlike loans guaranteed by government agencies, a conventional mortgage can be obtained fairly quickly. The institution has only to check the borrower's credit and job security, appraise the value of the property, and have the survey[1] and title[2] search brought up to date.

By law, private institutional lenders can only lend a certain percentage of the appraised value of the property. This is 75 to 80 percent of the value in most states. There are, however, some ways in which more than this percentage may be borrowed. Some insurance companies and others have been able to lend up to 90 percent of the appraised value of the property by issuing two mortgage notes which run simultaneously and carry the same terms as to interest and amortization.

The conventional loan path should be thought of seriously by every prospective homeowner. It is a business arrangement between a borrower and lender in which property value, credit rating, and mutual confidence play a dominant role—as they do in all normal business dealings.

## 3. CLOSING COSTS

These costs normally are paid by the borrower although in some cases, especially in V.A. loans, they may be absorbed by the lender or added to the amount of the mortgage. They consist of attorney's fees for searching the title and the prepara-

---

[1] A survey is performed by a licensed surveyor and shows the size, shape, and boundary markers of the property. It must be brought up-to-date to be sure that nothing has happened to the property since the last survey. It protects the owner as well as the lender.

[2] The bank's attorney searches or traces the history of the property to be sure that the property is free and clear of any claims against it. (The search is made in the land records of the locality where the property is located.) These records record any claims against the property, any easements or privileges granted to any person to use the property, and any faults in the clear title.

MORTGAGES 185

tion of the legal documents; any service charge by the lender; recording charges, and so forth. In F.H.A. mortgages the seller may be required to pay a discount charge which is the difference between the face value of the mortgage and the amount for which it can be sold by the lender to another institution.[3] (*see* Chapter 16, Section 2F, for further discussion of closing costs.)

## 4. TITLE INSURANCE

In some states the lender may require a Certificate of Title Insurance which guarantees the title and is issued by a title insurance company. Such a certificate or policy normally costs ½ percent of the face value of the mortgage. The reason for such a policy is that in many states the chain of ownership of a piece of property goes back so far that it is sometimes very difficult to trace it. This is where the title policy comes in. It is a certificate or policy issued by a title guarantee company which promises to indemnify an owner against any loss due to any defect in the title which may have occurred *prior* to the issuance of the policy. Such defects may be in the nature of liens, rights of way, or other encumbrances.

## 5. TRUTH IN LENDING

As of July 1, 1969, lenders are required by law to disclose to borrowers the annual percentage rate charged on a mortgage

---

[3] There is a market for F.H.A. and V.A. loans whereby the original lender can sell his mortgage to other institutions. The face value of the mortgage is not always the price. Because of low interest rates or other reasons, the seller may not be able to obtain the full face value of the mortgage.

loan to finance the purchase of residential real estate. This is especially valuable to the borrower when the lender requires him to pay "points" for obtaining the loan.

As mentioned in Section 2 of this chapter, during the past several years when very little mortgage money has been available, lenders, in order to increase their percent of interest, have discounted their loans in advance (in the example quoted, the real interest on a 7.5 percent loan would actually be 7.7 percent. The lender must tell you this or anything else about your loan which would increase your annual interest payments.

# CHAPTER 16

## BASIC REAL ESTATE LAW

1. What You Should Know About It
2. Common Legal Terms and What They Mean
   A. The Contract
   B. The Mortgage
   C. The Deed
   D. Easements
   E. Restrictive Covenants
   F. The Closing
3. Brokers

## 1. WHAT YOU SHOULD KNOW ABOUT IT

The house buyer, as well as the owner, should be familiar with a few of the common terms in real estate law and how they affect his rights and ownership. When you purchase or sell your house you will require an attorney to prepare and examine the necessary legal papers. The mortgagee will be represented by an attorney and the seller or buyer of your house well may be represented by an attorney also, and you should protect your rights by engaging an attorney of your own.

Even when you are represented you should know exactly what some of the terms used in real estate mean. The terms "easement" or "restrictive covenant" may be explained to you by your attorney; but if you are in the midst of a closing (to be explained later in this chapter) and your mind is not fully on the subject, you may discover later that you have purchased something you don't want.

## 2. COMMON LEGAL TERMS AND WHAT THEY MEAN

### A. THE CONTRACT

When you have decided to purchase property the first legal step is to enter into a contract. This may also be called the purchase agreement or the sales agreement, and it must contain all the essential agreements between the buyer and seller. These include the naming of the parties to the agreement; a description of the property to be sold; the agreed price; a date upon which the final sale is to take place; any matters regarding the title which are to be clarified or which now affect it; and a consideration which may be ten percent of the sales price.

BASIC REAL ESTATE LAW 189

B. THE MORTGAGE

The meaning of a mortgage has been explained in Chapter 15 but it bears repeating. A mortgage is a written document or conveyance (which is the legal term) of land (and property upon it) given by a mortgagor or debtor to a mortgagee or creditor as security for the repayment of the money loaned to the debtor to enable him to purchase the property. The mortgage becomes null and void upon the repayment of the debt. Be sure that you are familiar with all the terms and conditions of the mortgage and the obligations that it places upon you. If you are not sure, ask your attorney to explain it. For instance, you may want to know if you may be allowed to pay off the mortgage at any time without penalty.

C. THE DEED

When title to land is to be transferred from one party to another it must be done in a form prescribed by the law. This form is usually called a deed and must contain certain specific information so that it may be entered into the land records of the community where the property is located. The deed is the document that you will receive at the closing. It will convey all the title of the seller or grantor to you. It must contain certain essential information such as the names of the seller (grantor) and buyer (grantee); the consideration involved; the description of the land, including boundary lines and survey markers (if any) or reference to a plat[1] and a lot number; any exceptions or reservations (easements, restrictions, and so forth); words of conveyance which state that the grantor "conveys and war-

---

[1] A plat is a land map showing building lots.

rants" the property to the grantee;[2] and finally, the execution of the deed, which includes the proper signatures and the proper notarization.

## D. EASEMENTS

An easement is the right of someone other than the owner to enjoy certain privileges pertaining to the land. If you purchase property with a reservation such as an attached easement, it means that you are granting a right which is irrevocable to someone to use a portion of your land for his purposes.[3] The most common form of easement is the "right of way" which grants someone the right to cross your land by means of a defined road, to place a high-tension power line or an underground high-pressure gas line across your land. It may even involve a defined bridle path granted to the local riding or hunt club. *Be sure that any easement that may go with the land does not diminish your right to enjoy your land or your plans for any future expansion.* If you wish to oblige someone without permanently giving up your rights, you can grant him a "license" which can be revoked by you on short notice. (Please also note Chapter 3, Section 2G on Riparian Rights.)

## E. RESTRICTIVE COVENANTS

Such covenants are also known as "Deed Restrictions". For instance, it may be a condition of the deed that you must build a certain kind of house or use the property for only certain purposes. In many instances the developer of a new area will

---

[2] There are other legal words of conveyance. These are just a sample.
[3] Easements are always recorded in land records.

record the restrictions when he records the entire plot or development. In such a case it may be that your deed will state that your property is subject to the recorded restrictions. You must know what these restrictions are before you buy. In most cases the restrictions are contained in the body of the deed.

The purpose of most restrictive covenants is to provide a uniformity of construction style or size of house within a certain area. (These restrictions may be over and above those imposed by the zoning code.) There are also racial restrictive covenants which have to a large extent been invalidated by the law.

A restrictive covenant in your deed or as recorded can limit the kind of house you can build. If the restriction is on a house it can tell you what you can and cannot do in your house and even to whom you may sell it. Be careful, therefore, to note well any exceptions.

## F. THE CLOSING

The closing usually refers to the mortgage closing. As a final step in the purchase of your land or house, the closing is the event at which money is passed and the mortgage papers are signed. The lender or mortgagee gives you a check which you give to the seller of the property. He in turn gives you the deed to the property. In addition to the mortgage money you will also have to pay the difference between the selling price and the amount of the mortgage. This may be very little if it is a V.A.- or an F.H.A.-guaranteed mortgage.

In addition to the balance of the payment there will be several other charges you will have to pay. Some of these are:

> 1) The closing cost or fee, which is the cost to the lender of the appraisal of the property plus an attorney's fee for drawing up the papers.

2) The cost of a survey to show you your property lines and to make sure that the land or house you are buying is where you think it is.

3) Title insurance is required by lenders in many states and must be paid by you.

4) You must pay the fee to record your title in the land records of the community where your property is located.

5) You must, of course, furnish a fire insurance policy to the mortgagee. Such a policy should protect you against a number of hazards and is known as a homeowner's policy.

6) You will have to pay your share of the taxes from the date of your purchase to the next tax payment date. If your tax payments are to be collected monthly by your mortgagee (together with your mortgage interest and amortization) you will make payment to him.

7) In some states there is a mortgage tax which is based on the amount of the mortgage and which is paid by the purchaser.

8) If the previous owner has left fuel oil in the tank or bottled gas or other fuel or supplies that you will need, you will have to purchase these from him.

9) Finally, there is your own attorney to pay.

The total out-of-pocket expenses—which you pay only once—may run from one or two to several hundred dollars.[4] The other expenses at the closing, such as taxes, fuel cost, and insurance, are the ones you will have to pay from now on.

## 3. BROKERS

The real estate broker is an agent usually representing the seller of the property from whom in ordinary circumstances he obtains his commission. All states require that real estate brokers and their salesmen be licensed and most states have stringent laws about what a broker can and cannot do. It should be remembered by you, as the possible purchaser of a house (or land), that the broker usually can only submit your offer to the owner and has no authority to enter into a contract with you. If you pay money to a broker and if he should not remit it to the principal (owner), you have lost it. Therefore, deal only with reputable brokers and be sure they have the authority to receive and transmit an offer to the owner.

Brokers' commission rates are usually set by local real estate boards. If you ever wish to sell, be sure that you know about the various types of listing (the exclusive listing or the open listing); and do not accept an offer nor money from a broker unless you are sure you wish to sell at the price offered and to the people who are offering it.

---

[4]Closing costs may vary considerably among different lenders. Make inquiries at several banks. Efforts are being made by federal and state authorities to reduce closing costs.

# CHAPTER 17

## LANDSCAPING

1. A Definition
2. How Do You Want to Live Outdoors?
3. Making a Plan
4. The Details of Outdoor Space Planning
   A. Planning a Small Area
   B. Planning a Large Area
5. Climate and Landscaping
6. Fences

## 1. A DEFINITION

When you set out to look for a building site, one of the most important things to keep in mind is the way you want to live in your new house and the things you want to see around you. The ideal spot may exist only in your mind, but with luck you may find a stream or a hillside with a magnificent tree or wooded lot (complete with glade) that will tell you that *this is it*. Even if you purchase a lot in a subdivision you will be looking for a slope, a view, or an exposure that appeals to you. So it follows that as you choose the design and layout of your house, you should start to think of the design and layout of your land.

*The purpose of landscaping and planning your land is to make your outdoor environment as pleasant and livable as your indoor environment.*

## 2. HOW DO YOU WANT TO LIVE OUTDOORS?

In many parts of the country it is possible to carry on outdoor activities all year round. Even in some of the northern parts, one can spend three or four months in pleasant outdoor activities. In clement weather some families spend every possible moment outdoors while others are content to sit on a screened porch. Before you plan your outdoor environment, you must decide how you want to live in it by asking questions of yourself and your family:

Do you like to eat outdoors? All meals or just some meals?

Do you like outdoor cooking?

Do you like gardening? Vegetables, flowers, herbs, a rock garden, roses?

Do you have outdoor hobbies other than gardening? What are they and how much space do they require? (*See* Chapter 18 on hobbies.)

# LANDSCAPING 197

Do you want to work outdoors? Writing, painting, typing, the children's homework?

Do you want to relax and read a book, sunbathe, or just loaf?

Do you want a play space for the children where they can be watched from a work area such as a kitchen or laundry?

The answers to these questions will help you to define your plan.

## 3. MAKING A PLAN

The plan for your outdoor living should start at the same time as your house plans. If you have retained an architect you should think of where you want to locate your outdoor activities as you look at his first schematic drawings. If you have purchased plans you may be able to have them revised to fit your scheme.

You can make your plan in several ways. The easiest one is to obtain several sheets of graph paper and to use the squares as units of dimension. Depending on the total size of your site, you can make each square represent a certain number of feet. Your other tools are a pencil and a straight edge. Start by drawing the outline of your property which you can obtain from your survey or plot plan. The next step is to show the location of roads and trees, stone walls, rocky outcrops, and other natural features which you wish to use as part of your plan. You can locate these by the use of a long measuring tape and by using the corners of your property as reference points. On a small lot, you may not have many natural features, but plot them anyway. Now locate your house on the land. Indicate where the windows and doors are located on the outline of the house so that they may be used as reference points for your various areas of activity.

You are now ready to start your actual layout. You have a list of your special interests and requirements which should indicate how many spaces you will need for all your various outdoor pursuits. Remember that there should be reasonably defined areas not only for leisure activities but also for work and utility purposes.

## 4. THE DETAILS OF OUTDOOR SPACE PLANNING

Chapter 3, Section 2D ("Topography"), refers to the natural characteristics of the land which you should consider when choosing a site for your home. Even if you are buying a small building lot in a subdivision and if you have a choice of several locations, you can consider sun direction, prevailing winds, slope of the land, and any other natural features. There may not be any trees because developers are great believers in the bulldozer, but trees can be planted later where you want them.

If you are looking for a larger building site (one-half acre or more) you can be more selective about the natural characteristics of the land and, of course, you have more freedom in locating the house. If the plot is large enough, the house can face any direction you wish, within the zoning ordinance, with reference to slopes, rocky outcrops, trees, a stream, or a pond.

### A. PLANNING A SMALL AREA

To illustrate the planning process we will take a small building lot as an example. It is quite possible that a cleverly laid out 60 by 100 foot lot can give you as much satisfaction as a one-acre or larger site.[1] (*see* Figure 17.) Plan your outdoor space just as you do your space for indoor living and start with a general breakdown of the required space:

---

[1] Many large and medium sized cities are now zoning for not less than 6000 square feet and smaller towns usually don't allow less than one-quarter acre per single family house.

FIGURE 17

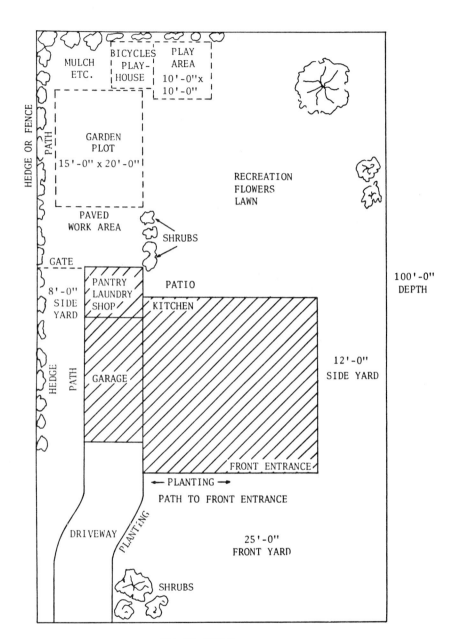

LAND PLAN FOR A SMALL PLOT

1) *A playground for the children.* This should be placed where it is readily accessible from your back door and where you can get a clear view of it from your kitchen window or from any door or window in your work area. Ten feet square is adequate for a sandbox, slide, and seesaw. You can leave adjacent space for a covered bicycle rack and a small playhouse. This play area need not be seen from anywhere but the back of your own house.

2) *A work area.* This is where you dry your laundry, place your garbage for collection, store your woodpile, barbecue wagon, and lawnmower. Of course this area can be kept neat, but even so it should not be in public view and you can tuck it away behind the garage and plant some tall shrubbery around it. It can be as small as twelve feet square and can be paved with loose flagstones, gravel, or crushed stone. Don't attempt to grow grass here. It would just be a nuisance to cut and it will be trampled constantly. The garbage man and the milk man and other delivery men should have free access through this area to your back door.

3) *A garden area.* The size of this area depends on your gardening enthusiasms. If you would like to grow some vegetables as well as flowers in rows for cutting, you require a minimum of fifteen feet by twenty feet and it should get the summer sun at least half the day. Be sure that you have a hose connection nearby for watering.

4) *A recreational area.* After setting aside space for working area, for play area, and for gardening, you have an area over forty feet square remaining. The number of uses to which you can put this space depends on your own desires. You can build a paved patio adjacent to the back door which leads from the kitchen, dining room, or living room. This will give you a permanent place for outdoor cooking or space for a movable cooker. If you want to, you can place an awning over part of the area to provide shelter against sun and rain. If you live in a cold climate with a short summer but like to be outdoors as often as possible, you can arrange a sheltered nook which will get the afternoon winter sun and still protect you from the

# LANDSCAPING

wind. Perhaps you enjoy sunbathing. Arrange a sheltered private spot for this.

The area beyond the patio can be any shape you like. Just remember to avoid corners—curves are much more appealing. This area can contain a small pool surrounded by flower beds. You can encircle your patio with raised planters. You even can install a play area for a sport which requires minimum space. (*See* Chapter 18, on hobbies.)

5) *The front yard.* If you use a typical zoning code requirement for a lot 60 by 100 feet, you must set the house back twenty-five feet from the front line (as well as a total of twenty feet from the side lines). This 25 by 60 foot area is your front yard and this is what everybody sees. Passers-by may never see your charming patio or private garden, so they will form their impression of you and the house only from the front.

The first consideration is the location of the garage and driveway. Unfortunately, there is not much that can be done with these on a narrow lot. Perhaps you can curve the driveway slightly so that you break the monotony of a row of houses each with a straight driveway. (A circular driveway, however, can look pretentious, or out of proportion, on all but the largest of lots.) If you curve it, you may be able to plant a clump of shrubs near the front property line. This planting can partially shield the garage from the street and thus prevent a possibly cluttered and never handsome garage interior from staring everyone in the face.

Your driveway can serve as a walk to the front of the house, but in that case you have to provide a footpath from it to the front door. Flagstones, brick, or any other paving material that suits your fancy may be used.

Foundation planting around a house requires care. Ask your local nurseryman or plant dealer about evergreen shrubs that don't grow too tall and that require only an occasional pruning to keep the house from being submerged. Window boxes or raised planters around the front door and flower beds around the driveway and walk add immeasurably to the appearance. If you don't want to be bothered with flower beds, you can use perennial flowering shrubs.

Finally we come to trees. If there are trees already on the property you can plan around them. If not, you can plant such ornamental trees as holly, dogwood, locust, redbud, flowering crabapple, and such tall growing shrubs as flowering quince or forsythia. Try to plant the taller growing material at the ends of the house in order to frame it. There is a world of possibilities and many good landscaping books to suggest them.

### B. PLANNING A LARGE AREA

Planning the various areas of activity for a large piece of land is essentially the same as for a smaller area. The difference is that you can allow more space for each activity and you can arrange these areas with relation to the house and the property boundary lines much more freely than you can a small plot. Read Section 2D of Chapter 3 carefully and answer the questions at the end of the chapter to get some direction.

A large plot of land is also likely to have more variety in elevation and other topographical features. You can, therefore, lay out your areas to take advantage of slopes, large trees, or outcrops of rock. There will be room enough to have a sweeping lawn, a rock garden, or a rose garden.

If your inclination is for a sedentary life, it is possible to keep the natural features of the land and to enjoy them without much gardening or lawn. In any case, don't feel obligated to use every square foot of your land for something. A *piece of meadowland, a grove of trees, or a bare hilltop left in its natural condition can be very appealing.*

After all, you are planning your outdoor living to suit your own family's life style and the only people you must please are yourselves.

## 5. CLIMATE AND LANDSCAPING

The climate of this country varies from the sharply contrasting seasons and sufficient rainfall of the North and Northeast through the semi-arid West and the arid Southwest to the subtropical South. Your choice of plants and your methods of planting must be in harmony with the climate. Instead of evergreen foundation planting you may have to use plants that are adapted to arid regions and to plant them in pots or planter boxes so that they can be easily watered. Consult local nurserymen and reliable books on gardening and landscaping.

## 6. FENCES

Fences can be of woven wood, saplings wired together, redwood, latticed for climbing plants, translucent to look like Japanese screens, low or high. Fences should be used only for a definite purpose such as to insure privacy for a dining or sunning area, to keep track of young children or pets, to provide a windbreak, to act as a boundary for a planted area, or to provide protection for a swimming pool.

Many communities do not allow fences to be built over a certain height—such fences (no matter what their actual purpose) are called spite fences.

## CHAPTER 18

# ROOM FOR RECREATION, HOBBIES, AND SPORTS

1. What Are Your Interests?
   A. Indoor Interests
   B. Outdoor Interests
2. The Actual Planning
   A. For Indoor Activities
   B. For Outdoor Activities

## 1. WHAT ARE YOUR INTERESTS?

The continuing trend in business and industry is toward reducing the number of hours that people spend at work. This, of course, means more time for you and you wonder how you can spend this leisure in a way that will bring relaxation, pleasure, and even profit.

It follows, therefore, when you are looking for a house to buy or build, that you keep in mind how you want to spend your leisure time and plan accordingly. Although this chapter cannot list all the leisure activities, it will mention the more common ones and give some information on how, with proper planning, you can provide for them.

As you read the following list of interests, you can check the ones that appeal to you enough to plan for them.

### A. INDOOR INTERESTS

- Collecting stamps or coins, bottles, rock samples, antique guns, wood carvings, and so on.
- Painting in oils or watercolor. This can be an indoor or outdoor activity but is usually both.
- Ceramics.
- Cooking. This does not mean cooking strictly to feed your family although this is included. Witness the popularity of cookbooks!
- Sewing. (Making things—not mending.)
- Reading or researching.
- Fishing or hunting. (Though far from indoor activities, these usually involve uses of indoor space for storage.)

Ham radio operations.

Card playing. Bridge parties.

Television watching.

Workshop for woodwork, metal, car tinkering, and so on.

Photography.

Sauna.

Exercising.

Games in general.

Tropical fish and other pets.

B. OUTDOOR INTERESTS

Archery.

Badminton.

Bowling—lawn, bocci, boules.

Croquet.

Gardening: rock garden, roses, vegetables, greenhouse.

Horseshoe pitching.

Shuffleboard.

Swimming pool.

Tennis—regular, paddle, or deck.

2. THE ACTUAL PLANNING

All of the aforementioned activities can be planned for when you build or buy, and many of them require no more than a little forethought. Chapter 10 ("Architecture and the Livable House") discusses room arrangements which will make it more convenient for you to pursue your outdoor or indoor leisure time activities. Remember that many such activities can take place in rooms that are also used for other purposes. Chapter 17 ("Landscaping") also discusses exterior planning for recreation or sports.

A. FOR INDOOR ACTIVITIES

1) *For the collector:* If you collect small objects you may require only a table with good light and a safe place to store your albums. Extensive shelves or glass-front cabinets for display or a section of blank wall for mounting objects may be desirable. You should arrange to have electrical outlets available for lighting your collection or your work space. While you are building or altering, you may want to have a cabinet made that will be suitable for displaying your collection or locking it in.

2) *Painting:* Look for a room that faces north and that is not shaded. See if you can arrange a clear space in the room to place an easel over a drop cloth. A cabinet or drawer space where you can store your materials will be convenient.

3) *Ceramics:* For ceramics, you should plan a space where you can "sling" clay without danger to rugs or furniture. A potter's wheel can create a good deal of slop. You must also arrange to have sufficient electrical current available for a kiln.

4) *Cooking:* In Chapter 11 ("The Heart of Your House") the subject of kitchens was discussed. If you require special utensils, chopping blocks, or storage space for particular foods, or more than one oven or refrigerator, be sure that your architect knows about it. If you are buying a completed house be certain you have enough space and extra electrical outlets in the kitchen.

5) *Sewing:* You require good light and a small clear space. Many people use a spare bedroom or bedroom corner.

6) *Reading or conducting research:* You will need adequate space, comfortable chairs, good light, a desk area to your

liking, built-in bookshelves, and possibly cabinets for keeping research books and notes safe.

7) *Fishing or hunting:* The inveterate fisherman or hunter may need room for rod racks, gun cases, trophy or tackle storage areas—even bookshelves for those favorite books and magazines. Of course, fly tiers or weapons finishers will want a special space to do their thing.

8) *Ham radio operation:* This activity requires a source of more than normal electrical energy and very often a tall pole for the antenna. This must be planned for. There must be spare electrical circuits, and the location of the antenna support, which usually has to be guyed, should be considered with reference to any outdoor sport or hobby that requires space.

9) *Card playing:* It may seem a waste of time to write about this, but if you like to play cards and do so frequently, a cabinet or closet into which you can slide card tables and store folding chairs is a bonus. Good light is essential and you can plan your outlets accordingly.

10) *Television watching:* A comfortable chair or sofa away from the children's homework or other activity and near the refrigerator is desirable. If you live far from a large city you may require a very high or a revolving antenna in order to receive more than one or two stations. Again you must plan for this with reference to other things you want to do.

11) *Workshop:* To most people the term "workshop" means a place where one makes or repairs things. The size and location of such an area depends, of course, on what your interest is. You should lay out a workshop area just as you would any work area in your house. Allow sufficient storage space for materials and for using power tools such as saws or lathes. Be sure you have provided for sufficient light and electrical outlets. If the shop is in an accessory building (such as a garage) you may have to provide heat. If it is located in the house, provide access to the work area so that lumber and metal can be transported to your shop and waste material carried out without damaging furniture or rugs. (*See* Figures 18 and 19.)

When you have laid out an adequate work area for the shop,

FIGURE 18

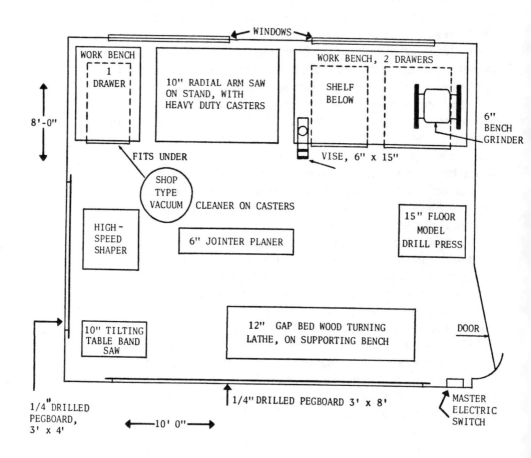

AN EXAMPLE OF AN 8'-0 x 10'-0" WORKSHOP
(AS LAID OUT BY A WOODWORKING HOBBYIST)

FIGURE 19

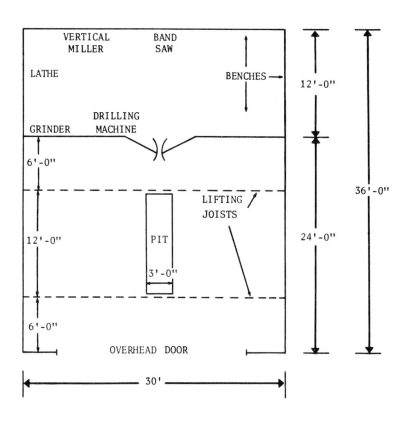

A SHOP FOR AUTOMOBILE REPAIR
AND RESTORATION

you must now fit it into the house. This may be the time for judicious cutting. Decide on essentials and make a list.

12) *Photography:* If you are interested in photography, you will know the equipment you need. You must arrange for a room that can be completely darkened and provide plumbing for sinks and electric current for printers, enlargers, and so forth.

13) *Sauna:* The use of the sauna has become a very popular recreation. The sauna is an import from Finland where it is always constructed outdoors, preferably on the shore of a stream or lake. In this country it is possible to purchase a prefabricated sauna room which can be assembled indoors in a basement area near a shower or outdoors near water. Saunas range in size from three feet by four feet to eight feet by twelve feet and even larger. The height varies from six feet six inches to seven feet, in proportion to the other dimensions. If you are thinking of building a sauna, you must provide a source of power. The approach to an outside sauna should be private.

14) *Exercise Area:* With everyone so fitness- and figure-conscious, an indoor area set aside for exercise machines and activities might well be appreciated. Some aspects to which you might give thought are mats or special flooring (if calisthenics, karate, and similar activities are a regular regimen), storage for accessories (e.g., weights), and permanent fixtures (bars or horses). The creation of a special exercise area could do a lot for those of us who have fallen into the habit of not exercising.

15) *Indoor games:* Such active games as billiards or table tennis can be provided for simply by supplying sufficient light and space for them. A billiard table can vary from four feet by eight feet to five feet by ten feet. Table tennis requires a table five feet by nine feet and thirty inches high. Allow at least seven feet six inches of clear height.

16) *Tropical fish and other pets:* As anyone who has ever gotten involved will know, tropical fish can become a passion requiring lots of space for tanks and equipment; electrical considerations also must be borne in mind. A strong interest in any type of pet or animal (especially if breeding is one of your

objectives), from dogs to hamsters to horses, may merit inclusion in your plans.

## B. FOR OUTDOOR ACTIVITIES

The areas required for the most popular outdoor family games are given below. The actual layout of such areas, the equipment, and the preparation of the surfaces can be found in books on these games.

Archery—A small target 12 inches in diameter can be used for a 150 foot range. This is not for competitive purposes where the target and range are much larger and longer. This activity should not be undertaken lightly. A pointed or even a blunt arrow can cause serious injury.

Badminton—20 feet wide by 44 feet long
Bowling, Lawn—20 feet wide by 120 feet long
Boccie—18 feet wide by 62 feet long (minimum)
Boules (a French version of boccie)—use a one-to-three ratio such as 12 feet wide by 36 feet long.
Croquet—30 feet wide by 60 feet long
Horseshoe pitching—10 feet wide by 40 feet long
Shuffleboard—6 feet wide by 52 feet long
Tennis, Doubles Court—36 feet wide by 78 feet long
    Paddle—20 feet wide by 44 feet long
    Deck—12 feet wide by 40 feet long

Some of the sports mentioned do not have firmly established measurements, but the ones given are the most commonly used.

Chapter 17 on landscaping discusses gardening and shows an example of the layout of a small plot. The enthusiastic gardener will know or be able to find out about soil, space requirements, cold frames, greenhouses, and rock gardens. The purpose of this chapter is merely to call your attention to the need for planning ahead all your leisure activities.

# CHAPTER 19

# MAINTAINING YOUR HOUSE

1. Choosing the Right Materials
   A. If You Are Building
      1) Excavation and Foundations
      2) Framing
      3) Masonry and Insulation
      4) The Exterior
      5) The Roof
      6) The Electrical System
      7) Plumbing
      8) Heating and Air Conditioning
      9) Flooring
      10) Cabinet Work
      11) Walls and Ceilings
      12) Painting
      13) Hardware
   B. If You Are Altering
2. What You Can Do About Maintenance
   A. Painting and Paperhanging
   B. Floors
   C. Plaster Patching
   D. Plumbing Repairs
   E. Electrical Repairs
   F. Basement Waterproofing
   G. Roof Repairs
   H. Miscellaneous

## 1. CHOOSING THE RIGHT MATERIALS

The physical maintenance of your house can be simple and inexpensive if the house was planned properly and if the right materials were used in its construction. With the present high cost of repairs, it has become almost a necessity for most families to "do it themselves." It is essential, therefore, to start with long-lasting and maintenance-free materials. When you are buying a completed house you can examine it carefully to see if you are getting these materials. It is when you are altering a house, building a new one, or buying one already under construction that you are able to choose the better materials. This chapter will emphasize the selection of the proper materials and methods of construction so that you may avoid having to make repairs for as long as possible.

### A. IF YOU ARE BUILDING

Chapter 13, "Building Your Own House," discussed planning and other matters concerning the making of a house to suit your mode of living. It stated that regular meetings with the architect are extremely important—among other reasons—for determining the proper materials and construction. If you don't have an architect you can discuss this matter with your builder or any knowledgeable person you know or have employed to look after your interests. If you buy a set of plans and specifications, you will notice that there are many blank spaces in the specifications for choices of materials, which you can fill in after you have received *expert advice*. Chapter 13, Section 6A, also discusses inspection of the construction during its important stages. Sound construction, of course, automatically eliminates many future maintenance problems, but additional precautions should be taken at certain critical stages.

# MAINTAINING YOUR HOUSE

1. Excavation and Foundations

If you have reason to suspect a wet basement, it will not cost very much to have the outside of your foundation wall covered with trowelled-on mastic waterproofing or hot pitch and felt paper. You also can have clay drain pipe laid in crushed stone at the bottom of your footings and have the water led to a dry well. Water always seeks the easiest way.

If you have heard about termites anywhere in the area, consider termite shields, which are laid on the foundation wall just under the wood sills.

2. Framing

It is always a good idea to have the sills impregnated with a good wood preservative. Be sure that the sills are securely bolted to the foundation walls, particularly where there are strong prevailing winds. A house can shift slightly under constant pressure of winds.

When you study the working drawings which show the sizes of the floor joists and the supporting beams or lintels over the doors and windows, it would be wise to examine the local building code. This will give the sizes of the beams for the various spans. Use the next larger size of beam if the sizes shown are marginal. (If there is any doubt ask the local building inspector.) This will prevent the cracking of ceilings and of the areas around doors and windows.

3. Masonry and Insulation

A properly laid exterior brick wall or other masonry wall should be almost maintenance free. If the brick wall is the cavity type (*See* Chapter 13, Figure 16) or if the masonry wall is furred on the inside, it will be weathertight as well.

Insulation should be of the blanket type which has been securely attached to the studs, rafters, or joists. Loose insulation settles after a while and loses its effectiveness. Be sure the insulation is vermin-proof and nonorganic.

The fireplace and chimney are part of the brick masonry work. The fireplace design is really the architect's job but you would do well to ask about it. A fireplace that draws well should be built to a certain proportion or relationship of width to height to depth. It should also have an efficient damper that will enable you to close off the fireplace chimney so that it does not create a draft when it is not in use. Try to make the front hearth as large as possible so that you can store wood on it and keep the fireplace ash and sparks away from rugs.

4. The Exterior

If you are building or buying a wood house you must be sure that wood is not located within several inches of bare earth. Wood in direct contact with earth invites termites, carpenter ants, and rot.

If you are planning to use a wood shingle exterior wall, consider predipped shingles. Impregnated shingles are much less likely to curl than others that are treated on only one side. Although it may seem like heresy to some purists, it is suggested that you look into metal siding. This siding is aluminum with a baked-on enamel finish and looks like painted clapboard. It is completely maintenance free and can be washed off with a hose.

Be sure that your windows and doors are painted with one coat as soon as they are installed—or even before. Window and door frames should be painted completely—the concealed portions as well as those that show. Get the best windows you can afford and have your exterior doors weatherstripped and at least 1¾ inches thick. This investment will pay for itself many times over in heating and cooling costs.

5. The Roof

If you are building or re-roofing with asphalt shingles, use heavier ones than the usual 210 pounds. Heavy butt shingles will last longer and are not so apt to lift off the roof in a heavy wind. If you can possibly afford it, use copper for flashing,

# MAINTAINING YOUR HOUSE

leaders, and gutters. It will last the life of the house. If you are using wood shingles, consider the predipped ones.

6. The Electrical System

Be sure that there are adequate spare circuits in the house so that you can add rooms or provide for air conditioners, extra appliances, flood lights, or any other future electrical needs. Ask your architect, builder, or electrical contractor for advice.

7. Plumbing

Use copper water lines. There are some sections of the country where we are told that this is not necessary. This may be true, but remember that copper lasts forever. Buy ceramic basins because enameled iron basins will not last. The finish will crack and rust. Use a cast iron enameled tub. If you have a hot water storage tank, use copper. Even a tiny hole in the glass lining of a noncopper tank will cause rust and leaks. If your plumber or builder tells you that an instantaneous hot water heater in your furnace will give you all the hot water you need—be skeptical. Two adults and two children taking morning showers will put a heavy strain on the heater. Have your plumber leave capped outlets in water and soil lines for future expansion of the house.

8. Heating and Air Conditioning

Instruct your architect or builder to get an oversize furnace and air conditioning unit. Usually the price of the next size is not much more and a unit that does not have to run at full capacity lasts much longer. Use natural gas or oil as fuel and get an annual maintenance contract from your fuel supplier to cover cleaning, service calls, and parts replacement. Heat pumps for heating and cooling are fine but few people can afford the cost of the electricity.

Consider hot water instead of hot air heating unless you are combining it with central air conditioning. If you are planning to expand your house, have the plumber or sheet metal man install capped outlets in the main heating loop, whether hot water or hot air, so that it doesn't have to be taken apart later. If you are planning to use underfloor gas heaters, be sure that they are well insulated from any wood.

9. Flooring

If the floors are oak, finish them with filler and wax, which will make them much easier to maintain. Don't use asphalt tile in the kitchen because it is not resistant to grease. Use linoleum or vinyl tile. If you are splurging on your kitchen, consider quarry tile or waxed flagstone.

Even if you are using wall-to-wall carpeting, it is still advisable to put a finished floor under it. It is cheaper to do it while the house is being built. If you don't, you or a new owner will be more or less forced to use wall-to-wall carpet unless you want the expense of refinishing the floor.

10. Cabinet Work

You may be dazzled by all the fancy finishes and curlicues that are now available on kitchen cabinets. Because of constant handling, the finishes on wood cabinets require frequent hard cleaning which is not good for the finish. Consider metal cabinets or a hard gloss finish on wood cabinets.

Unless you are designing an authentic period house, you should consider natural stained solid panel wood interior doors. They are inexpensive, they don't catch dust, and they can be cleaned easily.

11. Walls and Ceilings

Most new houses are now finished with gypsum wallboard. It is adequate if securely nailed, of proper thickness, and skillfully joined. Walls and ceilings of plaster over gypsum or wire lath are by far the best for sturdiness and for their fireproof and soundproof qualities. They are, however, considerably more expensive than wallboard.

# MAINTAINING YOUR HOUSE

12. Painting

The traditional lead and oil based paints for interior and exterior work have been largely replaced by synthetics. Latex base paints are also being used for interior finishing and the modern latex paint is washable (*not scrubbable*). For kitchen and bathroom areas, however, you should use a semi-gloss oil based paint or a good interior synthetic. Several well-known paint manufacturers make excellent interior washable paints.

You should be very careful about the paint you use for protecting exterior wood. You can't go wrong with lead and oil, but there are other paints that are excellent under the proper conditions. The first coat of paint is especially important. The adherence of all future coats will depend on the adherence of the base coat to the wood. If it is not done properly it can result in costly paint scraping and burning. Ask your painting contractor, or if you are doing it yourself, ask the local paint dealer or a builder. Some formulations are better for different parts of the country. If you are near salt water, in a hot moist climate, or in an area where there is constant direct sunshine, you should ask about the paint that is best suited to those conditions.

13. Hardware

The finished hardware for a new house is always chosen by the owner. The builder is told to include a money allowance for this hardware in his bid. If you exceed this allowance you will have to pay the extra. When you choose your hardware try to get solid brass doorknobs. There are, of course, knobs of porcelain, glass, or chrome plate, but many of these materials are fragile and plated knobs deteriorate. Many people like black iron hardware and there are latch sets available in this material; be sure to choose solid wrought iron. Be especially careful to get sturdy rustproof hardware for exterior doors. The best lock set available for an exterior door is economical in the long run.

## B. IF YOU ARE ALTERING

The extent to which you can use maintenance-free materials depends, of course, on the extent of the alteration. There is not much you can do about the original structural framing or foundations except to inspect them for soundness. If you find, for instance, that the sills are rotted, you can replace them with ones that are impregnated with wood preservative, and while you are about it you may as well install termite shields under them. If the foundation walls are cracked or seem to have settled in places, be sure to have them leveled with cement mortar before you place the termite shields and the sills. If it is an old house, it may not be properly fire stopped. You must do this. (*See* Figure 20.)

If the roof must be replaced you should use either heavy asphalt shingle or predipped wood shingle. Use copper for any flashing, leaders, or gutters that have to be replaced. All the materials that are mentioned in Section A of this chapter can be used if you are replacing or adding to your electrical, heating, and plumbing systems, including oversize electrical panels or furnaces or air conditioning.

Be careful about repainting, especially the exterior. Be sure that the underlying paint is adhering firmly to the wood. Don't use latex base paint or any synthetic unless you make sure that it will adhere to the original paint. These paints are almost always better suited to interiors.

Make a complete survey before you start any alteration work and be sure that the new material will not require constant care.

## 2. WHAT YOU CAN DO ABOUT MAINTENANCE

There are some men (and some women) who can perform almost any necessary maintenance jobs. Provided one has the

Fig. 20

## FIRESTOPPING A WOOD-FRAMED HOUSE

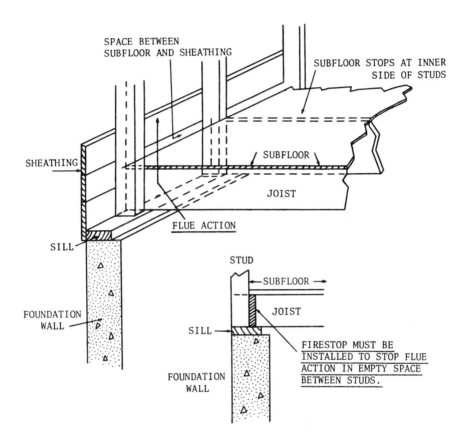

THIS IS AN OLD WAY OF FRAMING AND IT MUST BE FIRESTOPPED ON INSIDE OF STUDS.

MODERN FRAMING IS SHOWN IN CHAPTER 5 FIGURE 4

right tools and workshop facilities and the necessary skill, there are many jobs that can be done well by the amateur.

This section will confine itself to the simpler tasks which almost anyone can do—providing he has the time and the energy and wants to save the cost of hiring a professional.

### A. PAINTING AND PAPERHANGING

The one task that most people can do quite well is painting. With rollers on long handles for interior work, and latex or synthetic paints, you can do a job that is free of brushmarks and drips. The important thing is the preparation of the surface. Pay attention to the directions on the paint or patching plaster (or spackle) containers which will tell you about cutting out cracks and plastering them and about the proper preparation of the surface. Use good brushes. You can clean them and use them many times over.

For exterior painting, you must be prepared to stand on a high ladder or even on a sloping roof. You may have to remove loose putty from windows and reputty them. You will have to scrape and sand all loose paint. The author is old-fashioned enough to think that lead and oil exterior paint is best, but there are good acrylics and other synthetics. Be wary about using latex.

Because professional paperhangers are difficult to get and charge so much, many householders do it themselves. There are papers available with glue already applied so that all one has to do is to wet the paper and apply it to the wall. If you use a paper with edges which have to be trimmed and to which paste has to be applied, it would be wise to rent a paperhanger's table, knife, and straight edge.

The secret of successful paperhanging is the preparation of the surface to which the paper is to be applied. If it is a plaster wall in a new house, it must be thoroughly dry. If it is old paper it must be on securely. If it is not secure then you must remove it and prepare the wall for new paper. Use a size or sealer on the wall before you paper. There are special moistureproof wall coverings for bathrooms.

# MAINTAINING YOUR HOUSE

## B. FLOORS

Wood floors can be cleaned with steel wool and oxalic acid or a number of commercial cleaners. These will remove the wax finish and some of the surface dirt, but if this treatment is not sufficient to give you clean floors you will have to sand them. Sanders can be rented but must be used very carefully to prevent an uneven surface or gouges in the floor. The edges of the room must be scraped by hand or a small hand sander. Sanding floors creates a severe dust condition and the room being sanded must be carefully closed off. This is hard dusty work best left to professionals, but it can be done by the amateur.

Resilient flooring such as asphalt, vinyl asbestos, or vinyl tile are relatively simple to maintain if you are careful not to overwax. Use a light hand.

## C. PLASTER PATCHING

Most new houses have gypsum board walls and ceilings. Older houses have plaster over gypsum or wire lath or, in very o houses, plaster over wood lath. Plaster can crack and loosen because of settling of the structure, shrinkage of the wood framing, because it gets wet, and for many other reasons.

Mending a crack can be a comparatively simple operation. The crack must be carefully cut out and deepened so that the patching material will adhere properly. There are a number of patching compounds that can be used. Large cracks or areas of loose plaster require more care. The loose plaster must be removed, together with a margin of the sound plaster around it. Perhaps the lath underneath must be refastened. If you go back to the lath you must use a brown coat (or undercoat) to bring you to within less than a quarter of an inch of the finished surface. When the brown plaster is still slightly moist you can cover it with white plaster. All these materials have directions on their containers. With patience and care you can do a creditable job.

## D. PLUMBING REPAIRS

The amateur is advised not to attempt any plumbing repairs, beyond replacing a washer or cleaning out a plugged drain line, unless he is skilled and has the proper tools.

## E. ELECTRICAL REPAIRS

Even a rank amateur can wrap electrical tape around a potentially dangerous frayed iron or toaster cord, but repairing or adding to electrical circuits is not recommended for the amateur. it is difficult to collect insurance for a fire caused by electrical work that had not been done by a licensed electrician.

## F. BASEMENT WATERPROOFING

Basements which have been dry for years suddenly can become damp or wet because of heavy continuous rains, the changing of the direction of an underground watercourse, or the collapse of an old cistern, as well as for other reasons. The level of the water table may rise to about the level of the basement floor and exert hydrostatic pressure on the basement walls and floor. Often there is seepage through a weak point in the wall or at the point where the wall joins the floor. There are several waterproofing compounds available which can be mixed with cement mortar or are already mixed in a mortar. These materials, which can be brushed or troweled on the interior of the basement walls, will stop the seepage in the majority of cases. If the seepage occurs at the point where the walls meet the floor, you should apply a fillet of waterproof material after careful preparation to bond it securely to both floor and wall.

## G. ROOF REPAIRS

The amateur can repair or replace roof shingles which have been broken or torn away by high winds. Some people can even

# MAINTAINING YOUR HOUSE

repair defective flashing and re-solder and re-hang gutters and leaders. Do so if you have sufficient confidence in your own ability. The repair of flashing is especially tricky, however, and is not recommended for the weekend handyman.

## H. MISCELLANEOUS

You can try your hand at many other maintenance jobs such as installing bathroom accessories; installing folding closet doors; pointing up joints in masonry walls and chimney; repairing or replacing insulation. All it takes is time, proper materials and tools, and careful attention to the directions. There are many "how-to" books and magazines that are well worth reading.

CHAPTER **20**

# FIRE ALARMS AND BURGLAR PROOFING

1. Why?
2. Fire Alarms
3. Burglar Proofing
   A. Door Locks
   B. Window Locks
4. Burglar Alarms
5. Simple Precautions

1. WHY?

The wise householder gives serious thought to how quickly he may be aware of a fire in his home, especially when there are small children or older people involved. Personal safety entirely aside, a house fire can mean losses that no amount of insurance could ever restore. As for burglar proofing, it is unfortunate that a book about building, buying, and living in the best house you can should have to contain such information. Nevertheless, the casual breaking and entering of even small houses in remote places has become a way of life and it is foolish not to take some precautions. This chapter is not meant to alarm you, but simply to advise you on what is available for your protection *if you wish to use it.* Many people still keep their doors unlocked. If you are one of them, read this chapter and then ignore it!

2. FIRE ALARMS

It is recommended that everyone who lives in a nonfireproof wood house or a masonry house with wood framing have some kind of fire alarm in a central place.

For the single-family dwelling the simplest device is one that makes a loud noise when an electrical circuit is broken or a fusible link is melted by heat. A fusible link is simply a piece of wire that has a very low melting point. This wire holds a spring in tension. If the tension is released by the melting of the wire, the spring actuates a loud gong. These devices, which are quite reasonably priced, can be hung on the walls of central halls or stairway landings where any heat generated by a fire would go.

One simple electrical device depends on a wire strung around a room at the junction of the walls and ceiling. This wire has a very low melting point, and when it melts it actuates a relay

# FIRE ALARMS AND BURGLAR PROOFING

which causes a loud gong to ring. This gong can be located in the master bedroom.

There are many more elaborate alarms such as smoke detectors, pneumatic tubes, or thermostatic devices which depend on the smoke or heat of a fire to actuate them and which can be made extremely sensitive. These devices, however, are expensive and not usually necessary in the average sized house.

*It is important, however, that every household have a fire extinguisher suitable for electrical or grease fires.* Most of these contain a pressurized dry chemical which smothers the flame.

## 3. BURGLAR PROOFING

It is true that no lock or other device can keep a determined professional burglar from entering a house. This section, therefore, will be devoted to discussing the simple locks and other precautionary measures that can be taken to protect your house against the amateur or casual thief who may try to break in while you are away but who will be discouraged when he realizes that your house is securely locked. (Such devices also will make it more difficult for even an experienced thief.)

### A. DOOR LOCKS

When you are building or altering you must insist that your builder install good locks in your exterior doors. The mortise cylinder lock with a deadbolt is the best. This lock is set (mortised) into the door. It has a spring latch which you can fix so that the door can be opened from the outside without a key (when you are at home). The spring latch has one slanted face which faces outward so that when you shut the door the latch slides back to allow the door to close and then slides into the door frame to keep the door closed. This slanted face, which makes it so easy to close and lock the door witout a key, also makes it vulnerable to an intruder. When this latch is set to lock the door, even an amateur can force it open by slipping a piece of celluloid or metal between it and the door frame. What he cannot force open is a deadbolt—a heavy piece of metal with a square face that slides into the doorframe when the key is turned in the lock. Most of these locks have a small knob instead of a key on the inside that can be turned to open or close the deadbolt. If the door is part glass, however, an intruder, by breaking a single pane, can put his arm in and unlock the door by twisting this knob. To counteract this, the newer locks now contain a deadbolt that cannot be opened from either side except by the use of a key. In addition, some of these locks contain a noisy alarm or buzzer that goes off if someone tries to force the door. Cylinder locks are pickproof to all but an expert.

The door chain which allows you to partly open the door so that you can speak to someone at the door and still prevent his entry has long been a way of life for city dwellers and has now become a wise precaution for country and suburban dwellers as well.

The sliding glass patio or glassed porch door normally is fitted with a latch that swivels into the door frame to keep it closed. Unfortunately, this latch can be forced. You can fit your sliding doors with locking bolts that fasten to the frame or track. Any intruder will hesitate to smash a large sheet of double thick glass to gain entry.

## B. WINDOW LOCKS

The double hung window is almost always supplied with a window lock which swivels under a keeper so that neither portion of the window can be opened. This swivel lock is good and if it is closed *tightly* it cannot be forced open. An intruder can, however, break a pane of glass and open it from the outside. You can fit your windows with locking bolts to prevent this. You also can fit your windows with a device that allows them

# FIRE ALARMS AND BURGLAR PROOFING

to be opened wide enough for ventilation but not wide enough for an intruder to get in. In addition, sliding windows and casement windows can be fitted with locking bolts when they are fully closed, or some other devices to keep them partly open for ventilation.

## 4. BURGLAR ALARMS

An effective burglar alarm is usually too expensive for the normal householder. The bank safe deposit box is still the best place to keep your valuables.

For those interested, alarms can be of several kinds—the ultrasonic, which sets off an alarm when the intrusion of a body changes or interrupts sound wave patterns; photoelectric alarms, which work on the same principle as the electric eye door opener with which most of us are familiar; proximity alarms, which may be used to guard a bank vault or the Queen's jewels but are not commonly used in a private home.

The most economical type of alarm is one that makes a loud noise or turns on lights or both at any sign of forcible entry (but this can cost close to a thousand dollars). These devices generally are not connected to a central station. Any such connection costs a considerable sum annually and is more expensive than the television set or typewriter that the intruder might steal.

There are also window and door wiring devices which are actuated when a door or window is opened while the alarm is set. These alarms are mostly silent alarms—i.e., they set off an alarm at some central point after the intruder is in the house. The electrical household alarm should be a perimeter alarm which makes a loud noise and turns on lights *before* the intruder is in the house. There are companies that install such alarms at reasonable cost.

There is in existence now, Burglar Insurance, at quite reasonable rates, which are guaranteed by the Federal Government. Please ask your insurance agent about this.

## 5. SIMPLE PRECAUTIONS

Without getting elaborate or frightened, there are many simple precautions that you can take to protect your household against fire and intruders.

Simple spring fire alarms, fire extinguishers, and good locks on your doors are not expensive and, once installed, you can forget about them.

A few flood lights around the house which can be turned on from interior switches are also easy to install, especially at the time of building or altering.

One final suggestion is the electric clock switch which turns the lights on and off when you are away. Some people have two of these—one turns the living room lights on and the other turns bedroom lights on just after the living room lights are turned off.

Here we are almost at the end of the twentieth century and this is what some of us have to do. . . .

# index

Air conditioning, 219-30
Alarms
   burglar, 233
   fire, 230-31
Almost-new house, 78-79
   examination of, 81
Alteration, *see* Remodeling
American Institution of Architects, 153
Architects
   contract and duties of, 153-55
   fee paid to, 156-57
   planning meetings with, 155-56
Architecture, 102-22
   effect of climate and availability of materials on, 112-13
   livable layout and, 113-20
      room arrangement, 114-19
      room size, 119-20
   momentary and superficial styles in, 103-9
      cathedral ceiling, 104-6
      conversation pit, 104
      Norman turret and Tudor, 107
      "olde English" cottage, 107-9
      "phony" mansard roof, 106-7
      picture window, 103
      shutters, 106
      split level, 103-4
   regional, 110-11
   of townhouses, 38
Artificial lakes, 94-99
Attic, remodeling of, 145-46

Basement
   remodeling of, 145-46
   waterproofing of, 226
Bathroom
   in new house, 132-33
   planning of, 129-32
   remodeling, 128-129
Bedrooms, remodeling, 144-45
Bids for construction contract, 157-59
Brokers, real estate law and, 193
Brush fires, 30
Budget, 6
Builders
   of developments, 52-53
   of speculative houses, 68-76
Building a home, 148-69
   completion of, 168-69
   construction contract for, 157-60
      obtaining bids, 157-59
   inspection during construction in, 160-67
      important stages, 161-65
      progress, 165-67
   insurance during construction for, 168
   materials for, 216-21
      cabinet work, 220
      electrical system, 219
      excavation and foundations, 217
      exterior, 218
      flooring, 220
      framing, 217
      hardware, 221
      heating and airconditioning, 219-20
      masonry and insulation, 217-18
      painting, 221
      plumbing, 219
      roof, 218-19
      walls and ceilings, 220
   lot for, 148-50
      purchase of, 149-50
      topography of, 149
      zoning and, 148-49

Building a home (*continued*)
  payments during construction for, 167
  preliminary decisions on, 150-52
    financial considerations, 151-52
  working drawings for, 152-57
    choosing architect, 153-57
    purchased plans and builder's design, 152-53
Burglar proofing, 231-34

Cabinet work
  inspection of, 164-65
  materials for, 220
  payment during, 167
  progress of, 166
Carpentry, finish
  inspection of, 164-65
  payment during, 167
  progress of, 166
Cash position, 5
  building a home and, 151-52
  future, 6-7
Cathedral ceiling, 104-6
Ceilings, materials for, 220
Certificate of Title Insurance, 185
"Cinderella style," 109
Cities, 10
  renovated houses in, 22
Climate
  architecture and, 112-13
  landscaping and, 203
Closing
  costs of, 184-85
  legal definition of, 191-93
Cluster housing, 42
Complete new city, 12
Condominiums, 23, 39-42
  zoning and, 26
Congress, U.S., housing legislation in, 40, 181
Construction
  contract for, 157-60
  inspection during, 160-67
  of mobile homes, 47
Contract
  with architect, 153-55
  construction, 157-60

  legal definition of, 188
Contractor for remodeling, 139-40
Convenience
  of locality, 15-16
  room arrangement and, 114-18
Conventional mortgages, 183-84
Conversation pit, 104
Cooperatives, 23, 44-45
  zoning and, 26
Countryside, 11-12, 23-24
Covenants, restrictive, 190-91

Dams on artificial lakes, 97-98
Deed
  legal definition of, 189-90
  restrictions in, 190-91
Developments, 38, 50-65
  advantages and disadvantages of, 50-52
  on artificial lakes, 94-99
  builder of, 52-53
  future maintenance in, 63
  physical structure of houses in, 53-62
    building code, 59-61
    foundations, 54-55
    framing, 55-59
    inspection during construction, 54
    interior finish, 62
  prices in, 62-63
Do-it-yourself maintenance, 224-27
Door frames, installation of
  inspection of, 162
  progress on, 166
Door locks, 231-32

Easements, 190
Economic level of locality, 13
Electrical system
  installation of
    inspection of, 162-64
    payment during, 167
    progress on, 166
  materials for, 219
  of old house, rehabilitation of, 87-88
  repairing, 226

Elevation, 30
Embankments on artificial lakes, 97-98
Estimates on remodeling, 142-43
Excavation, 217
  inspection of, 161
  progress on, 165-66
Exercise area, 212
Exterior
  materials for, 218
  remodeling of, 141, 145-46
  of speculative houses, 70-72
Exurbs, 11-12

Family room, 144
Family size, 4
  future, 6
Federal Housing Administration, 40, 46
  mortgages available from, 179-81
  remodeling financed by, 146
Fee, architect's 156-57
Fences, 203
Financing
  of remodeling, 146
  *See also* Mortgages
Finish carpentry, 164-67
Finish work
  of development house, 62
  inspection of, 164
  progress on, 166
Fire alarms, 230-31
Floors
  finished, 164, 166
  maintenance of, 225
  materials for, 220
  in old houses, rehabilitation of, 88-89
Forest fires, 30
Foundation
  building code requirements for, 59-61
  of development houses, 54-55
  inspection of, 161-62
  progress on, 166
  materials for, 217
Framing
  building code requirements for, 61

  of development houses, 59-61
  inspection of, 162
  materials for, 217
  of old house, rehabilitation of, 86-87
  progress on, 166
Front yard, 201-2

Garden, 200
Gas mains, 27
G.I. loans, 181-83

Hardware, 221
Heating system
  installation of, 162-64, 166
  maintenance of, 219-20
  of old house, rehabilitation of, 87-88
Hobbies
  room arrangement and, 119
  *See also* Leisure time interests
Housing Act (1961), 40
Housing and Urban Development, Department of, 46, 47, 181

Inspection during construction
  when building a home, 160-67
  of development, 54
Insulation materials, 217-18
Insurance, 172-75
  burglar, 234
  during construction, 168
  title, 185
Interior
  of development houses, 62
  remodeling of, 140-41, 144-45
  of speculative houses, 72-73

Job prospects, 6

Kitchen, 124-28
  planning, for new house, 127-28
  remodeling, 124-25
  in speculative houses, 125-27

Lakes, artificial, 94-99
Landscaping, 196-203
  climate and, 203

Landscaping (*continued*)
   fences and, 203
   of large area, 202
   outdoor living and, 196-97
   of small area, 198-202
Landslides, 28
Laundry room, 133-35
Layout, livable, 113-20
Lease before purchase, 80-81
Leisure time interests, 206-13
   indoor, 206-7
   outdoor, 206
   planning space for, 208-13
Lie of the land, 28
Lifestyle, 3-4
   room arrangement and, 118-19
   size of house and, 4
Livable layout, 113-20
   room arrangement and, 114-19
   room size and, 119-20
Living room, remodeling of, 144
Local builders, speculative houses built by, 68-76
Locks
   on doors, 231-32
   on windows, 232-33
Lot, choice of, 148-50

Maintenance
   of development house, 63
   do-it-yourself, 224-27
      basement waterproofing, 226
      electrical repairs, 226
      floors, 225
      painting and paperhanging, 224
      plaster patching, 225
      plumbing repairs, 226
      roof repairs, 226-27
   materials and, 216-22
      building a home, 216-21
      remodeling, 222
Mansard roof, 106-7
Masonry, materials for, 217
Materials
   for alterations, 222
   availability of, 112-13
   for building a home, 216-21

Mobile Homes Manufacturers Association, 47
Mobile homes, 45-48
   construction of, 47
   in mobile home parks, 47-48
   zoning and, 26
Mode of living, *see* Lifestyle
Modular houses, 48
   zoning and, 26
Momentary architectural styles, 103-9
Mortgages, 178-86
   closing costs of, 184-85
   conventional, 183-84
   Federal Housing Administration, 179-81
   title insurance and, 185
   legal definition of, 189
   truth in lending and, 185-86
   Veterans' Administration, 181-83
Mudslides, 28-29

New houses, 38
   under temporary lease, 78
Norman turret style, 107
Not-so-old house, remodeling of, 140-42

"Olde English" cottage, 107-9
Older house, 42-44, 79
   examination and assessment of, 82-83
   rehabilitation of, 86-92
      heating, plumbing and electrical system, 87-88
      plaster, 89-90
      roof, 90-91
      structural frame, 86-87
      time and effort for, 91-92
      woodwork and floors, 88-89
   remodeling of, 139-40
   zoning and, 25
Outdoor living
   landscaping and, 196-97
   room arrangement and, 118-19
   space planning for, 198-202
      for games and sports, 213

Outdoor living (*continued*)
   space planning for (*continued*)
      for large area, 202
      for small area, 198-202

Painting
   do-it-yourself, 224
   inspection of, 165
   materials for, 221
Paperhanging, 224
Payments during construction, 167
Pets, planning space for, 212-13
Picture windows, 103
Planning
   of bathroom, 129-33
   of kitchen, 127-28
   meetings with architect for, 155-56
   for remodeling, 138
Plans, purchased, 152-53
Plaster
   in old house, rehabilitation of, 89-90
   patching cracked or broken, 225
Plastering, 164, 167
Playground area, 200
Plumbing
   materials for, 219
   installation of, 162-64, 166
   of old house, rehabilitation of, 87-88
   repairing, 226
Prevailing winds, 31
Purchased plans, 152-53

Real estate law, 188-93
   brokers and, 193
   closing in, 191-93
   contract in, 188
   deed in, 189-90
   easements in, 190
   mortgage in, 189
   restrictive covenants in, 190-91
Recreation area, outdoor, 200-1
Regional architecture, 110-11
Remodeling
   of exterior, 145-46
   financing of, 146
   of interior, 144-45
      bathroom, 128-29
      bedrooms, 144-45
      family room, 144
      kitchen, 124-25
      living room, 144
   of not-so-old house, 140-42
   obtaining estimates on, 142-43
   occupying house during, 143
   of old house, 139-40
   preliminary planning of, 138
Rentability, 36-37
Restrictive covenants, 190-91
Riparian rights, 31
Roof
   laying of, 162, 166
   materials for, 218-19
   of old house, rehabilitation of, 90-91
   "phony" mansard, 106-7
   repairing, 226-27
Rooms
   arrangement of, 114-19
   size of, 119-20
Rural areas, 11-12

Salability, 26-27
Sauna, 212
Schools, 14
   transportation to, 24-25
Sewer lines, 26-27
Shutters, 106
Site, quality of, 24-32
   prevailing winds, 31
   riparian rights, 31
   topography, 27-30
   transportation, 24-25
   utilities, 26
   zoning, 25-26
Size of house, 4
Slope of the land, 28
Speculative houses, 39, 68-76
   examination of, 70-73
   how to find, 68-69
   kitchen in, 125-27
   purchase of, 69-70

Speculative houses (*continued*)
    zoning of, 73, 76
Split-level houses, 103-4
Sports, planning space for, 213
Suburbs, 10-11, 22-23
Sun direction, 27-28
Superficial architectural styles, 103-9

Taxes, 14-15
Title insurance, 185
Town roads, 27
Townhouses, 23
    architecture and appearance of, 38
Topography, 27-30, 149
Transportation, 12-13
    to shopping and schools, 24-25
    to work, 24
Trees, 29-30
Truth in lending, 185-86
Tudor style houses, 107

Utility areas, 124-35

Vacation homes, 94-99

Veterans' Administration, mortgages available from, 181-83
Veterans' Housing Act (1970), 182

Wallboard, installation of, 164, 166
Walls, materials for, 220
Water lines, 26-27
Waterproofing of basement, 226
Window frames, installation of, 162, 166
Window shutters, 106
Windows
    locks on, 232-33
    picture, 103
Wiring, *see* Electrical system
Woodwork, 88-89
Work areas, 124-28, 133-35
    outdoor, 200
    remodeling of, 144-45
Working drawings, 152-57
Workshop, 209-12

Zoning, 16-18, 26
    choice of lot and, 148-49
    remodeling and, 141-42
    of speculative houses, 73, 76